Selected Titles in This Series

(Continued in the back of this publication)

Financial Markets
Stochastic Analysis and the
Pricing of Derivative Securities

Translations of
MATHEMATICAL
MONOGRAPHS

Volume 184

Financial Markets
Stochastic Analysis and the Pricing of Derivative Securities

A. V. Mel'nikov

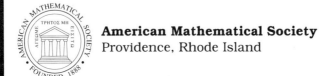

American Mathematical Society
Providence, Rhode Island

А. В. Мельников

ФИНАНСОВЫЕ РЫНКИ

«Научное издательство ТВП», Москва, 1997

Translated from the Russian by H. H. McFaden

1991 *Mathematics Subject Classification.* Primary 90–02, 90A12, 90A09;
Secondary 90A60, 60H30, 60G42, 60G40.

ABSTRACT. This is one of the first textbooks in the new subject of "actuarial and financial mathematics". The abandonment of a fixed gold price at the beginning of the 1970s and the opening of the Chicago market in the trading of option contracts, combined with the appearance of the famous papers of Black and Scholes and of Merton on pricing options, led to that rare situation when the opportunity for rapid practical utilization creates a favorable environment for purely theoretical developments (in the area of pricing derivative instruments on a financial market). Due to the powerful effect of contemporary mathematical methods and stochastic analysis, there was a transition from arithmetic to mathematics as a basis for finance. Financial mathematics not only has acquired the outlines of an independent science, but also has found effective and promising applications in financial and insurance markets. This is the direction of the current book, which is a self-contained and sufficiently broad introduction to the general mathematical theory of pricing for derivative securities.

Library of Congress Cataloging-in-Publication Data

Mel'nikov, A. V., 1953-
    [Finansovye rynki. English]
    Financial markets : stochastic analysis and the pricing of derivative securities / A. V. Mel'nikov ; [translated from the Russian by H. H. McFaden].
        p. cm. — (Translations of mathematical monographs; v. 184)
    Includes bibliographical references and index.
    ISBN 0-8218-1082-0 (alk. paper)
    1. Derivative securities–Prices–Mathematical models. 2. Options (Finance)–Prices–Mathematical models. I. Title. II. Series.
HG6024.A3M43813 1999   99-17735
332.64′5–dc21   CIP

Contents

Foreword

Financial mathematics is at present going through a period of intensive development, especially in the area connected with contemporary *stochastic analysis*. It is the methods of the general theory of random processes that have turned out to be most suitable for an adequate description of the evolution of *basic* (bonds and stocks) and *derivative* (forwards, futures, options, and so on) securities.

Historically, the first work (1900) in this direction was the dissertation of Bachelier [13], a student of Poincaré who, several years before Einstein and 23 years before Wiener, gave a mathematical definition of the concept of 'Brownian motion', used it to model the dynamics of stock prices, and gave a formula for the investment cost of an option. The main deficiency of Bachelier's model, which was the possible negativity of the stock prices, was removed in 1965 by the well-known economist Samuelson, who proposed a *geometric Brownian motion* for describing these prices. This model now bears the names of Black and Scholes, who in 1973 [15] obtained precise formulas for computing the fair price and hedging strategies for European options in the framework of the model.

Employing the heuristic argument that stock prices are either rising or falling at any moment of time, Cox, Ross, and Rubinstein [19] proposed regarding these changes as *discrete* and introduced a *binomial model* of a financial market. They showed that the binomial model has a geometric Brownian motion 'as a limit', and the formula obtained for a fair price converges to the Black–Scholes formula.

These now-classical papers have become a direct basis for the application and development of methods from contemporary stochastic analysis in the mathematical theory of finance. It is in this direction, with the use of elements of functional analysis and convex analysis, that deep results have been obtained about the structure of prices and about the properties of arbitrage and completeness of a financial market.

The goal of this book is to present, in a sufficiently self-contained form, the methods and results of the contemporary theory of financial computations for a discrete market. It gives a representation of basic techniques in stochastic analysis: martingales, semimartingales, stochastic exponents, Itô's formula, Girsanov's theorem, and so on. The discreteness of the models considered above leads to a whole series of technical simplifications, and often to greater clarity of the results obtained. Yet at the same time, this discrete theory contains in itself many elements of the very complex techniques and problems in the general theory. Therefore, the book can be regarded as a sufficiently broad introduction to the contemporary mathematics of financial computations with derivative securities.

In large part this book is based on the material and approaches expounded in [12], and it represents the content of the course of lectures "Stochastic analysis in finance" given by the author in 1994–1997 in the Mechanics and Mathematics

Department of Moscow State University. This explains its theoretical character and direction.

The author sincerely thanks A. N. Shiryaev, Yu. M. Kabanov, and D. O. Kramkov for many useful discussions and for their help, criticism, and constant support. The book was published at the proposal of the scientific publishing house "TVP". The author is very grateful to V. I. Khokhlov, both for making this proposal and for the amount of work he put into editing the book.

November 22, 1998   *A. V. Mel'nikov*

Notation

$(\Omega, \mathcal{F}, \mathbf{P})$	a probability space
$X, Y, \ldots$	random variables (RVs)
$\tau, \sigma, \ldots$	stopping times
$(\Omega, \mathcal{F}, \mathbb{F}, \mathbf{P})$	a stochastic base
$\mathcal{O}$	the optional σ-algebra
$\mathcal{P}$	the predictable σ-algebra
$\sigma\{\varepsilon_1, \ldots, \varepsilon_n\}$	the σ-algebra generated by the random variables $\varepsilon_1, \ldots, \varepsilon_n$
$\mathbb{P}^*$	the set of martingale measures
$[\![\tau, \sigma[\![$	a stochastic interval
$\mathbf{P}\{A\}$	the probability of an event A
$\mathbf{E}\,X$	the mathematical expectation of X
$\mathbf{D}\,X$	the variance of X
$\mathbf{E}\,(X \mid \mathcal{A})$	the conditional mathematical expectation of X with respect to the σ-algebra $\mathcal{A}$
$\mathcal{E}_n(U)$	the stochastic exponential with respect to a sequence U
$\langle M \rangle$	the quadratic characteristic (compensator) of a square-integrable martingale M
$\Phi(x)$	the standard normal distribution
$L_2(\Omega, \mathcal{F}, \mathbf{P})$	the set of random variables X such that $\mathbf{E}\,|X|^2 < \infty$
$\mathbf{P} \sim \mathbf{P}^*$	equivalence of two probability measures
$\mathbf{R}^d$	d-dimensional Euclidean space
$\mathbb{Z}_+$	the set $\{0, 1, \ldots\}$ of nonnegative numbers
$\mathbf{I}_A$	the indicator function of a set A
$\varnothing$	the empty set

ΔX_n $= X_n - X_{n-1}$

$a \wedge b$ $= \min\{a, b\}$

$a \vee b$ $= \max\{a, b\}$

a^+ $= a \vee 0$

$[a]$ the integer part of a number a

$f \sim g$ equivalence of functions f and g as $x \to a$, which means
                              that $\lim_{x \to a} f(x)/g(x) = 1$

$o(x)$ a function such that $\lim_{x \to 0} o(x)/x = 0$

$\binom{N}{n}$ the number of combination of N things, taken n at a time

X^π the value of a strategy (portfolio) π

$\mathbb{C}(N)$ the fair price of an option with expiration time N

$\mathbb{C}_T(\Delta)$ the fair price of an option wth expiration time T and dis-
                              creteness step Δ

(f, N) a contingent claim

(B, S)-market   a market made up of the assets B and S

SF   the set of self-financing strategies

$\mathrm{SF}_{\mathrm{arb}}$ the set of arbitrage strategies

$\Pi(x, f, N)$ the set of hedging strategies for (f, N) with initial value x

GF   the set of G-financing strategies

LECTURE 1

> The first lecture bears an introductory, mainly nonmathematical, character and is intended to give a general notion of a financial market, its structure and organization, and its basic and derivative instruments. The corresponding material can be found in [1], [7], [11], [17], [18], [21], [23], [24], [31].

Basic concepts and objects of a financial market

§ 1. Financial markets

A *market* is a multilevel concept. Here the term is used as the totality of demand, supply, and realization of assets at the prices arising on this basis. The main object of our attention is a *financial (stock) market*, which can be separated into a *securities market (financial market)* realized on a stock exchange (assets, bonds, and derivative securities), and a *financial resources market outside the stock market* (bank services, loans, credits, derivative instruments in turnovers outside the stock market).

Participants in a financial market are banks, financial companies, insurance companies, and other financial-insurance structures, including individuals. The forms of engagement in financial activities are very diverse — from buying, selling, owning, and borrowing securities to receiving dividends and allocating capital to direct consumption.

Under contemporary conditions a qualified business requires sufficiently accurate pricing of the assets being traded on the market. Such pricing is not possible without a certain "idealization" of the market. The concepts have thereby arisen of a "*frictionless*" *market* (all operations are realized instantaneously and without cost), a *highly liquid market* (the possibility of instantaneous purchase and sale of assets), and so on. Assets realized in terms of securities consititute the basis of a financial market. They include, first and foremost, stocks and bonds.

§ 2. Basic securities

Stocks are *share securities* issued with the goal of accumulating capital for subsequent activities. The owner of a stock, or the stockholder, gets both the right to participate in the control of the company (according to the rule: the number of shares is equal to the number of votes), and to receive dividends. To control the

1

company the stockholders hire a staff of employees, and one of the main tasks of the personnel hired is to achieve an increase (to the maximal possible degree) in the price of the shares.

Bonds are *debt securities* issued by a government or by various types of firms with the goal of accumulating capital, restructuring its debts, and so on. In contrast to stocks, they are issued for a certain time period, at the expiration of which they are removed from circulation by *repayment* (redemption). Characteristics of a bond are the following: exercise (redemption) time, redemption cost (face value), payments up to redemption (coupons). An essential indicator of the "quality" of a bond is its *return up to the redemption time* (yield). The yield allows us to compare payments for bonds with different characteristics and is, in essence, equivalent to a bank *interest rate*.

As an example we consider a bond with face value $A = 100,000$ (dollars, rubles), circulation time $N = 10$ years, annual 8% coupons $K = 0.08A = 8,000$ (dollars, rubles). Suppose that at a given time O it is sold at the price $B(O, N) = 101,000$. Then the corresponding interest rate r (and hence also the yield of the bond) can be determined from the relation

$$B(O, N) = \sum_{n=1}^{N} \frac{K}{(1+r)^n} + \frac{A}{(1+r)^N}$$

and is equal to 7.85%, that is, its yield differs from the coupon payments. It is now clear from this example that the yield of a bond (the interest rate) is a function of the sale time and the redemption time. Therefore, investigation of the *term structure of interest rates* (yield) is one of the most important problems for financial operations with bonds.

As a rule, a bond is a *nonrisky asset* in comparison with stocks, whose prices can be fairly chaotic. Nevertheless, buying such securities also involves *the risk that the stipulated promissory notes will not be honored*. As a rule, bonds with a higher degree of risk also pay a higher return. Least risky are government bonds, which, consequently, have the least return.

Upon the issue of a series of bonds *the possibility of early redemption* is also taken into account. In this case the owner of the bond is paid a *premium* on the scale of, say, one year's return, and the company which issued such a bond insures itself against the necessity of paying high interest when the interest rate falls. The return on such bonds is higher than that on ordinary bonds.

§3. Derivative instruments of a financial market

Stocks and bonds are *primary*, since they are determined directly in terms of economic factors. In contrast to them, *derivative (secondary) securities* "function" on the basis of the *basic* securities (stocks and bonds) already at hand on the market.

The attractiveness of the derivative securities market is due to the fact that it requires essentially smaller initial expenditures than basic securities. A consequence of this is that investors with "meager" purses take part, and this increases the intensity of the bargains made and the liquidity of the market. Along with the game of buying and selling securities with the goal of extracting profits (*speculation*), derivative instruments are attractive also from the point of view of *insurance against loss*.

Forward transactions. Company A plans to buy shares of company B at the end of the year. Insuring against a possible increase in the price of a share of B, company A concludes a forward transaction with B. According to the transaction, A is obligated to buy the shares of B at a fixed and stipulated (forward) price F. In Figure 1 we show the payment of A in dependence on the price S_N of the shares of B.

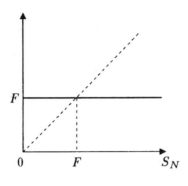

FIG. 1

Options. Company A has acquired shares of company B because it hopes for a return from owning them. However, the shares might fall in price, and to insure against such a decrease company A buys a *seller's option*, that is, the right to sell the shares to B at the end of a year at a fixed price K. Figure 2 shows the total payments to the holder of the shares and of the option at the end of a year.

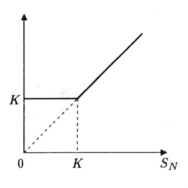

FIG. 2

In contrast to a forward transaction, to buy an option company A must pay a *premium*, which can be a considerable amount. The exact calculation of this premium, called the *price*, is a difficult and very urgent problem in options trading.

Among other secondary securities we mention the following:

a *warrant* is a security issued by a firm in exchange for cash and giving its owner the right to buy a fixed number of shares in this firm at a fixed price at any time up to a fixed date;

a *convertible bond* is a bond allowing its owner, at any time up to redemption, to exchange it for a specific number of shares;

caps, collars, and floors are derivative instruments allowing a "diversification" of types of loans, imposing lower, upper, or two-sided bounds on the (floating) rate.

An important novelty on the financial market (outside the stock exchange) is a *swap*: a private agreement between two parties about an exchange of money flows (possibly in different currencies) at a definite time in the future according to a formula stipulated in advance.

§ 4. Activities of an investor on a financial market

Calling anything of value that can be sold, bought, or exchanged an *asset*, we have defined a financial market to be a *collection of such assets*. Then it is natural to define an *investor* to be a participant in a financial market investing free capital in various assets, and to call the collection of those assets his investment *portfolio*. The real art of an investor is the ability to regularly and dynamically build his portfolio of investments (*to manage his portfolio*). The simplest management techniques include such activities as holding onto an asset, buying it, selling it, and borrowing it. The redistribution of a portfolio can have the goal of limiting or entirely eliminating the risk in various financial transactions (for example, the purchase or sale of an option). In this case one speaks of *hedging* or protecting one's investments, and the corresponding dynamic portfolio is called a *hedging portfolio*.

As a rule, carrying out the operations indicated above requires of the investor certain expenditures called *operating costs*. They include the commission fees of a broker, stock exchange, clearing house, and so on. In general the chain of commission payments is more essential for a small investor, since the large corporations, who have their own networks of dealers on the stock exchanges, are already freed from a large part of the commissions, and for the large investors these constitute fractions of a percent of the volume of transactions completed. Therefore, in a number of cases it is possible to neglect the operating costs.

For an investor it is essential to find an investment strategy that gives a profit with zero initial expenditures. Such strategies are called *arbitrage* strategies. We explain the concept of arbitrage by the following simple example.

Suppose that an asset sells in Frankfurt at a price of 150 DM, and in New York at a price of $100. The current exchange rate of marks versus dollars is 1.55. In this case the following chain of operations is possible:

> borrow 150 DM;
> buy the asset in Frankfurt for 150 DM;
> sell it in New York for $100;
> exchange the $100 for 155 DM at the exchange rate;
> repay the borrowed 150 DM.

The result is a clean profit of 5 DM, and hence an *arbitrage strategy* has been constructed.

Another example of arbitrage, specifically Russian, was given by the trading of *vouchers*, or privatized checks. This monetary authorizing document was a certain "generalized" stock and was the most liquid security on the Russian stock market in 1993. Trading of vouchers was implemented on a number of stock exchanges. Analyzing the results of these trades on the RCRE (Russian Commodity and Resource Exchange, Moscow) and the SPSE (St. Petersburg Stock Exchange), one could see a fairly large difference in prices. For example, from May 20, 1993 to May 26, 1993 the "fork" of prices (to within tens) between St. Petersburg and Moscow was at least 0.1 thousand rubles, reaching its maximum at 0.6 thousand rubles on May 21, 1993. (5.8 thousand rubles – 5.2 thousand rubles). This created the possibility of the following natural sequence of actions by the participants in the vouchers market:

> buy checks in Moscow, sell them in St. Petersburg, and receive a riskless return of 100 to 600 rubles from each voucher.

To realize this *arbitrage opportunity*, use was made of a network of dealers with large banking supplies of vouchers in the localities, which removed the problem of transferring large batches of vouchers in a short time (twenty-four hours) and hence reduced the *operating expenses*.

It is clear that arbitrage cannot exist over a long period, because the actions of the very same investors who want to exploit it lead to the vanishing of the arbitrage opportunity.

§ 5. Interest rates and discount rates

An important element of the operation of a stock market is a *loan*, that is, capital borrowed or lent. The standard procedure for obtaining a loan is described by the following two quantities: the time N of the loan, and the interest rate $R(N)$, which is equal to the client's return over the time N per unit of capital lent. Consequently, the final amount of capital $B(N)$ is equal to $B(O)(1+R(N))$, where $B(O)$ is the initial capital lent.

To compare the returns for loans with different loan times the sizes of the interest rates are recomputed on the basis of one year:

$$r = r(N) = R(N) \frac{\text{number of days in a year}}{N}.$$

Here one says that $r = r(N)$ is the *yearly interest rate* computed annually, semi-annually, quarterly, ..., if N is a year, half-year, quarter, ... An example of such a yearly interest rate is the well-known rate LIBOR (London Interbank Offered Rate).

Another characteristic of a loan is the *discount rate*. A bank can attract capital over a time N by announcing an interest rate $R(N)$ yielding a return of $B(O)R(N)$, or by announcing a discount rate $Q(N)$ giving a client the opportunity to obtain the sum $B(N)$ by an initial investment of the sum $(1-Q(N))B(N)$ in the bank. While in the first case the mechanism for receiving a return is realized by simply opening a *bank account*, in the second case it is realized by issuing a bond with face value A that is sold with discount $Q(N)A$, and hence the current price $B(O, N)$ of the bond is equal to $(1 - Q(N))A$.

The discount rates are recomputed on an annual basis according to the formula

$$q = q(N) = Q(N) \frac{\text{number of days in a year}}{N} = Q(N)\,m.$$

It is not hard to see that all these quantities are connected by the relation

$$\left(1 + R(N)\right)\left(1 - Q(N)\right) = \left(1 + \frac{r}{m}\right)\left(1 - \frac{q}{m}\right) = 1,$$

which permits them to be computed.

§ 6. Stock exchanges, clearing houses, and interbank markets

A special organization called a *stock exchange* has been created to manage the trading of spot goods, futures contracts, options, and other financial instruments, and to provide informational and other services. It is a nonprofit organization that finances its activities by collecting membership fees and fees for services offered. Its base is made up of the *stock exchange members*, who are given so-called *brokers' seats* on the stock exchange. The value of a broker's seat lies in the opportunity of engaging in trading activities directly in the hall of the stock exchange and with smaller commission fees. The members of a stock exchange actively making use of their trading privileges include *brokers*, special agents who "bring parties together" to close contracts.

As a rule, the trading of futures contracts is implemented through a *clearing house*, a nonprofit organization that "relieves" clients of the necessity of checking out opposite parties. In a clearing house special accounts (*a margin*) are opened for the parties in a contract, their positions are registered, and the margin is backed up, that is, the margin is taken to the necessary level. The clearing house has a guarantee fund at its disposal to cover expenses. As a rule, members of the clearing house are also members of the stock exchange.

Most international currency activities take place on the *interbank market*, a global network of financial institutions connected by brokers and a telecommunications service that operates around the clock. The main geographical centers are found in New York, London, and Tokyo. There are three groups of participants in the interbank market: makers, brokers, and clients.

The most active banks operate as makers who establish the prices of purchases and sales and who buy and sell on demand. The clients engage in currency operations according to the prices set by the makers.

§ 7. Basics of a mathematical model of the dynamics of prices on a financial market

The statistics of securities prices leads one to deduce a "random character" for their dynamics. Therefore, it is possible to get an adequate description of the evolution of prices by identifying a model and its statistical regularities, which "are controlled" by an unknown probability distribution $\mathbf{P}$.

The impossibility of identifying the measure $\mathbf{P}$ exactly implies the impossibility of an exact probabilistic description of the behavior of the possible events. However, practice shows that market experts usually do not disagree about whether this or that event A will take place. There is divergence of opinion only about the numerical value of $\mathbf{P}\{A\}$, which will be either strictly positive for all the experts or zero for all the experts. It is thus reasonable to propose as the initial object of the model not simply a probability space $(\Omega, \mathcal{F}, \mathbf{P})$, but a *family of probability spaces* $(\Omega, \mathcal{F}, \{\mathbf{P}\})$, where $\{\mathbf{P}\}$ is a class of mutually *equivalent* probability measures, each of which either does or does not give weight to the possible events A: $\mathbf{P}\{A\} > 0$ for all $\mathbf{P} \in \{\mathbf{P}\}$ or $\mathbf{P}\{A\} = 0$ for all $\mathbf{P} \in \{\mathbf{P}\}$.

An inherent feature of the contemporary mathematics of finance is the active use of *stochastic analysis*, both for describing the dynamics of asset prices and for pricing diverse instruments of a financial market. The second lecture is devoted to an exposition, based on [2], [9], and [10], of the elements of "discrete stochastic analysis" with such concepts as martingales, semimartingales, stochastic integrals, stochastic exponentials, Itô's formula, and so on. The apparatus of stochastic exponentials will be systematically used in the subsequent lectures (see [2], [4], [5]).

The elements of discrete stochastic analysis

§ 1. A stochastic base

A discrete *stochastic base* $(\Omega, \mathcal{F}, \mathbb{F}, \mathbf{P})$ is defined to be a probability space $(\Omega, \mathcal{F}, \mathbf{P})$ with a *filtration* $\mathbb{F} = (\mathcal{F}_n)_{n \in \mathbb{Z}_+}$, that is, a nondecreasing collection of sub-σ-algebras $\mathcal{F}_n \subset \mathcal{F}$.

A filtration $\mathbb{F}$ determines on a probability space a certain evolutionary and informational structure, which in turn determines a way of looking at random processes as functions of two variables defined on $\Omega \times \mathbb{Z}_+$.

All the subsets of $\Omega \times \mathbb{Z}_+$ are called *random sets*.

A random variable $\tau \colon \Omega \to \mathbb{Z}_+ \cup \{\infty\}$ is usually called a *stopping time* (a Markov time, or a time "not depending on the future") if $\{\tau \leq n\} \in \mathcal{F}_n$ for any $n \in \mathbb{Z}_+$.

We remark that this condition is equivalent to the requirement that $\{\tau = n\} \in \mathcal{F}_n$.

Let $\mathcal{F}_\infty = \sigma\{\bigcup_{n \in \mathbb{Z}_+} \mathcal{F}_n\}$ be the σ-algebra generated by $(\mathcal{F}_n)_{n \in \mathbb{Z}_+}$. We define the σ-algebra $\mathcal{F}_\tau$ of events "preceding" the stopping time τ:

$$\mathcal{F}_\tau = \{A \in \mathcal{F}_\infty \colon \quad \{\tau \leq n\} \cap A \in \mathcal{F}_n, \quad n \in \mathbb{Z}_+\}.$$

Among all the sequences $(X_n)_{n \in \mathbb{Z}_+}$ of random variables we distinguish those for which the X_n are $\mathcal{F}_n$-measurable (denoted by $X_n \in \mathcal{F}_n$, $n \in \mathbb{Z}_+$). Such sequences adapted to the filtration $\mathbb{F}$ are called *stochastic sequences*.

For stopping times $\tau \leq \sigma$ we define the random sets

$$[\![\tau, \sigma[\![= \big\{(\omega, n) \colon \tau(\omega) \leq n < \sigma(\omega)\big\},$$
$$[\![\tau, \sigma]\!] = \big\{(\omega, n) \colon \tau(\omega) \leq n \leq \sigma(\omega)\big\},$$

which are called *stochastic intervals*. The stochastic interval $[\![\tau]\!]$ is called the graph of τ. Let $[\![O_A]\!] = A \times O$ for $A \in \mathcal{F}_0$.

The filtration $\mathbb{F}$ induces in the space $\Omega \times \mathbb{Z}_+$ the *optional* σ-algebra $\mathcal{O}$ and the *predictable* σ-algebra $\mathcal{P}$ as the σ-algebras generated by all the stochastic intervals of the forms $[\![0, \tau[\![$ and $[\![0, \tau]\!]$, $[\![O_A]\!]$, respectively.

A stochastic sequence $(X_n)_{n \in \mathbb{Z}_+}$ is said to be *predictable* if $X_n \in \mathcal{F}_{n-1}$ for each $n \in \mathbb{Z}_+$, where $\mathcal{F}_{-1} = \mathcal{F}_0$. It turns out that each predictable sequence is measurable (as a function of two variables) with respect to $\mathcal{P}$, and each stochastic sequence is measurable with respect to $\mathcal{O}$.

For a set $A \in \mathcal{O}$ the random variable

$$\tau_A(\omega) = \begin{cases} \inf\{n \in \mathbb{Z}_+ : (\omega, n) \in A\} & \text{if} \quad \{\cdot\} \neq \varnothing, \\ \infty & \text{if} \quad \{\cdot\} = \varnothing \end{cases}$$

is defined and is called the *debut* of A. The debut of A is a stopping time; moreover, τ_A is a *predictable* stopping time for $A \in \mathcal{P}$ in the sense that

$$\{\tau_A = 0\} \in \mathcal{F}_0, \quad \{\tau_A = n\} \in \mathcal{F}_{n-1}, \qquad n \in \mathbb{Z}_+.$$

In particular, the *first hitting time*

$$\tau_B^X = \inf\{n \in \mathbb{Z}_+ : X_n \in B\},$$

of a Borel set B by a stochastic sequence $(X_n)_{n \in \mathbb{Z}_+}$ is the debut of the optional set $\{(\omega, n) : X_n(\omega) \in B\}$, and hence it is a stopping time (a predictable stopping time for a predictable sequence).

For a stochastic sequence $(X_n)_{n \in \mathbb{Z}_+}$ and a stopping time τ we define

$$X_\tau(\omega) = \sum_{n=0}^{\infty} X_n \mathbf{I}_{\{\tau=n\}}(\omega) + X_\infty \mathbf{I}_{\{\tau=\infty\}}(\omega),$$

where X_∞ is a random variable that is measurable with respect to $\mathcal{F}_\infty$ (for example, $X_\infty \equiv 0$ or $X_\infty = \lim_n X_n$ if this limit exists), and we define the *stopped sequence* $X_n^\tau = X_{n \wedge \tau}$. It follows immediately from the definition that $X_\tau \in \mathcal{F}_\tau$, and a stopped sequence is a stochastic sequence (predictable if the original $(X_n)_{n \in \mathbb{Z}_+}$ is predictable and τ is a predictable stopping time).

§ 2. Martingales

Along with predictable sequences, *martingales* have an important place in stochastic calculus and analysis.

DEFINITION 2.1. A stochastic sequence $(X_n)_{n \in \mathbb{Z}_+}$ is called a *martingale* if $\mathbf{E}|X_n| < \infty$ and $\mathbf{E}(X_n \mid \mathcal{F}_{n-1}) = X_{n-1}$ (a.s.) for all $n \in \mathbb{Z}_+$.

An integrable stochastic sequence $(X_n)_{n \in \mathbb{Z}_+}$ is called a *submartingale* (*supermartingale*) if (a.s.)

$$\mathbf{E}(X_n \mid \mathcal{F}_{n-1}) \geq X_{n-1} \quad (\mathbf{E}(X_n \mid \mathcal{F}_{n-1}) \leq X_{n-1}).$$

We note that from a martingale $(X_n)_{n \in \mathbb{Z}_+}$ it is easy to construct a submartingale $Y_n = g(X_n)$, where $g = g(x)$ is a convex function.

EXAMPLE 2.1. Let $(\eta_n)_{n\in\mathbb{Z}_+}$ be a sequence of independent Bernoulli random variables with

$$\mathbf{P}\{\eta_n = 1\} = p, \quad \mathbf{P}\{\eta_n = -1\} = q, \qquad p + q = 1.$$

Interpreting $\eta_n = \pm 1$ as a "success" or "failure" of a player in the nth game, we define the predictable (with respect to the *natural* filtration $\mathcal{F}_n = \sigma\{\eta_0, \ldots, \eta_n\}$) stochastic sequence $V_n = V_n(\eta_0, \ldots, \eta_{n-1})$ to be the bet of the player in the nth game. An example of such a strategy is the *martingale strategy*, when the bets are doubled upon losing, and the game is stopped upon winning, that is, $V_n = 2^{n-1}\mathbf{I}\{\eta_1 = -1, \ldots, \eta_{n-1} = -1\}$. The total payoff in n games is

$$X_n = X_{n-1} + V_n\,\eta_n = \sum_{k=0}^{n} V_k\,\eta_k.$$

The "average" change

$$\mathbf{E}\left(X_n - X_{n-1}\,|\,\mathcal{F}_{n-1}\right) = \mathbf{E}\left(\Delta X_n\,|\,\mathcal{F}_{n-1}\right)$$

in the total payoff at the nth step is zero (a fair game), or nonnegative (a favorable game), or nonpositive (an unfavorable game). It is clear that in the first case (X_n) is a martingale ($p = q = 1/2$), in the second case it is a submartingale ($p \geq q$), and in the third it is a supermartingale ($p \leq q$).

EXAMPLE 2.2. Let $X = (X_n, \mathcal{F}_n)_{n=0,\ldots,N}$ be a stochastic sequence such that $\mathbf{E}|X_n| < \infty$ and $\mathbf{E}\,X_\tau = \mathbf{E}\,X_0$ for any stopping time τ. Then X is a martingale.

Indeed, for $n \leq N$ and a set $A \in \mathcal{F}_n$ we define the stopping time

$$n_A = n\,\mathbf{I}_A + N\,\mathbf{I}_{\Omega\setminus A}.$$

Then $\mathbf{E}\,X_N = \mathbf{E}\,X_0 = \mathbf{E}\,X_{n_A} = \mathbf{E}\,X_n\,\mathbf{I}_A + \mathbf{E}\,X_N\,\mathbf{I}_{\Omega\setminus A}$. Consequently, $\mathbf{E}\,X_N\,\mathbf{I}_A = \mathbf{E}\,X_n\,\mathbf{I}_A$, and hence $\mathbf{E}\left(X_N\,|\,\mathcal{F}_n\right) = X_n$.

EXAMPLE 2.3. A whole series of examples of martingales (sub- and super-martingales) can be obtained from the above *procedure for stopping* a stochastic sequence $(X_n)_{n\in\mathbb{Z}_+}$ with the help of a stopping time τ.

If $X = (X_n)_{n\in\mathbb{Z}_+}$ is a martingale (sub-, supermartingale), then so is $X^\tau = (X_{n\wedge\tau}, \mathcal{F}_n)_{n\in\mathbb{Z}_+}$.

Since

$$X_{n\wedge\tau} = \sum_{k=0}^{n-1} X_k\,\mathbf{I}_{\{\tau=k\}} + X_n\,\mathbf{I}_{\{\tau\geq n\}},$$

it follows that $X_{n\wedge\tau} \in \mathcal{F}_n$ and $\mathbf{E}|X_{n\wedge\tau}| < \infty$.

Further, $X_{(n+1)\wedge\tau} - X_{n\wedge\tau} = \mathbf{I}_{\{\tau>n\}}(X_{n+1} - X_n)$ and $\mathbf{E}\left(X_{(n+1)\wedge\tau} - X_{n\wedge\tau}\,|\,\mathcal{F}_n\right) = \mathbf{I}_{\{\tau>n\}}\mathbf{E}\left(X_{n+1} - X_n\,|\,\mathcal{F}_n\right) = 0$ (in the case of a martingale X).

EXAMPLE 2.4. Let $\mathbf{P}^*$ be another measure given on the discrete stochastic base $(\Omega, \mathcal{F}, \mathbb{F}, \mathbf{P})$. We denote these measures by $\mathbf{P}_n$ and $\mathbf{P}_n^*$ if they are regarded on the σ-algebra $\mathcal{F}_n$ ($\mathbf{P}_n$ and $\mathbf{P}_n^*$ are the "restrictions" of the original measures to $\mathcal{F}_n$).

The measure $\mathbf{P}^*$ is said to be *absolutely continuous* with respect to $\mathbf{P}$ ($\mathbf{P}^* \ll \mathbf{P}$) if $\mathbf{P}\{A\} = 0 \implies \mathbf{P}^*\{A\} = 0$ for any $A \in \mathcal{F}$. In this case a nonnegative random variable Z called the *density* is defined and denoted by $Z = d\mathbf{P}^*/d\mathbf{P}$. If also $Z > 0$, then the measures are said to be *equivalent* ($\mathbf{P} \sim \mathbf{P}^*$).

In the case when $\mathbf{P}_n^* \ll \mathbf{P}_n$ or $\mathbf{P}_n^* \sim \mathbf{P}_n$ one speaks of the *local absolute continuity* $(\mathbf{P}^* \overset{\text{loc}}{\ll} \mathbf{P})$ or the *local equivalence* $(\mathbf{P}^* \overset{\text{loc}}{\sim} \mathbf{P})$ of these measures, and the variable

$$Z_n = d\mathbf{P}_n^* / d\mathbf{P}_n$$

is called the *local density*.

If $A \in \mathcal{F}_n$, then for the local density we have that

$$\mathbf{E}\, Z_{n+1}\, \mathbf{I}_A = \mathbf{E}\, \frac{d\mathbf{P}_{n+1}^*}{d\mathbf{P}_{n+1}}\, \mathbf{I}_A = \mathbf{P}_{n+1}^*\{A\}$$

$$= \mathbf{P}_n^*\{A\} = \mathbf{E}\, \frac{d\mathbf{P}_n^*}{d\mathbf{P}_n}\, \mathbf{I}_A = \mathbf{E}\, Z_n\, \mathbf{I}_A.$$

Consequently, $(Z_n, \mathcal{F}_n)_{n \in \mathbb{Z}_+}$ is a *martingale with respect to* $\mathbf{P}$. In the case when $\mathbf{P}^* \ll \mathbf{P}$, the local density Z_n and the density Z are connected by the relation $Z_n = \mathbf{E}(Z \mid \mathcal{F}_n)$, and $Z_n \to Z$ ($\mathbf{P}$-a.s.) as $n \to \infty$.

These last properties of the local density Z_n and the density Z are valid for the class of *uniformly integrable martingales* $(M_n, \mathcal{F}_n)_{n \in \mathbb{Z}_+}$:

$$\lim_{C \to \infty} \sup_n \mathbf{E}\, |M_n|\, \mathbf{I}_{\{|M_n| > C\}} = 0,$$

for which the limit $M_\infty = \lim_{n \to \infty} M_n$ always exists ($\mathbf{P}$-a.s.), $\lim_{n \to \infty} \mathbf{E}\, |M_n - M_\infty| = 0$, and $M_n = \mathbf{E}(M_\infty \mid \mathcal{F}_n)$.

EXAMPLE 2.5. An integrable stochastic sequence $\xi = (\xi_n, \mathcal{F}_n)_{n \in \mathbb{Z}_+}$ is called a *martingale-difference* if for any $n \in \mathbb{Z}_+$

$$\mathbf{E}(\xi_n \mid \mathcal{F}_{n-1}) = 0$$

($\mathbf{P}$-a.s.).

It is clear that partial summation of martingale-differences leads at once to a martingale (and conversely):

$$X_n = \sum_{k=0}^n \xi_k \text{ is a martingale} \quad \Longleftrightarrow \quad (\xi_k) \text{ is a martingale-difference}.$$

A stochastic sequence $A = (A_n, \mathcal{F}_n)_{n \in \mathbb{Z}_+}$ is said to be *increasing* if $\Delta A_n = A_n - A_{n-1} \geq 0$ (a.s.) for all $n \in \mathbb{Z}_+$. The following important theorem shows the close connection among the classes of submartingales, martingales, and predictable increasing sequences.

THEOREM 2.1 (Doob decomposition). *Let* $X = (X_n, \mathcal{F}_n)_{n \in \mathbb{Z}_+}$ *be a submartingale. Then there exist a martingale M and a predictable increasing sequence A such that for any* $n \in \mathbb{Z}_+$

(2.1) $$X_n = M_n + A_n$$

($\mathbf{P}$-*a.s.*). *This decomposition is unique.*

PROOF. Let $M_0 = X_0$, $A_0 = 0$, and

$$M_n = M_0 + \sum_{k=0}^{n-1} \left\{ X_{k+1} - \mathbf{E}\left(X_{k+1} \,|\, \mathcal{F}_k \right) \right\},$$

$$A_n = \sum_{k=0}^{n-1} \left\{ \mathbf{E}\left(X_{k+1} \,|\, \mathcal{F}_k \right) - X_k \right\}.$$

Clearly, M and A have the required properties of the decomposition (2.1).

If we assume the existence of another such decomposition with a martingale M' and a predictable increasing sequence A', then

$$\Delta A'_{n+1} = A'_{n+1} - A'_n = \Delta A_{n+1} + \Delta M_{n+1} - \Delta M'_{n+1}.$$

Since A and A' are predictable and M and M' are martingales,

$$\Delta A'_{n+1} = \mathbf{E}\left(\Delta A'_{n+1} \,|\, \mathcal{F}_n \right) = \mathbf{E}\left(\Delta A_{n+1} \,|\, \mathcal{F}_n \right) = \Delta A_{n+1},$$

and hence $A_n = A'_n$ and $M_n = M'_n$ for $n \in \mathbb{Z}_+$.

The theorem is proved.

A martingale $M = (M_n, \mathcal{F}_n)_{n \in \mathbb{Z}_+}$ with finite second moment $\mathbf{E} M_n^2 < \infty$ is said to be *square integrable*. Since M^2 is a submartingale, we have from Theorem 2.1 that

$$M_n^2 = m_n + \langle M \rangle_n,$$

where m is a martingale, and $\langle M \rangle$ is a predictable increasing sequence, which in this case is called the *quadratic characteristic* of M.

It is clear that

$$\langle M \rangle_n = \sum_{k=1}^{n} \mathbf{E}\big((\Delta M_k)^2 \,|\, \mathcal{F}_{k-1} \big),$$

$$\mathbf{E}\big((M_k - M_l)^2 \,|\, \mathcal{F}_l \big) = \mathbf{E}\left(M_k^2 - M_l^2 \,|\, \mathcal{F}_l \right)$$
$$= \mathbf{E}\big(\langle M \rangle_k - \langle M \rangle_l \,|\, \mathcal{F}_l \big) = 0, \qquad l \le k,$$

$$\mathbf{E} M_k^2 = \mathbf{E} \langle M \rangle_k, \qquad k \in \mathbb{Z}_+.$$

Further, if M and N are two square-integrable martingales, then their *mutual quadratic characteristic* (covariance)

$$\langle M, N \rangle_n = \frac{1}{4} \left\{ \langle M + N \rangle_n - \langle M - N \rangle_n \right\}$$

is defined, and the stochastic sequence $(M_n N_n - \langle M, N \rangle_n)_{n \in \mathbb{Z}_+}$ is a martingale.

Square-integrable martingales M and N are called *orthogonal* if $\langle M, N \rangle_n = 0$. Fixing M and considering all possible martingales N orthogonal to M, we can construct a whole family of square-integrable martingales of the form

$$X_n = M_n + N_n.$$

Conversely, any square-integrable martingale is representable in this form, called the *Kunita–Watanabe decomposition*.

Example 2.2 gave a simple characterization of the martingale property in terms of a stopping time. The same direction is involved in the following theorem of Doob (the optional sampling theorem).

THEOREM 2.2. *Suppose that the martingale (submartingale)* $(M_n, \mathcal{F}_n)_{n \in \mathbb{Z}_+}$ *and the stopping times* $\tau_1 \leq \tau_2$ *(a.s.) are such that* $\mathbf{E}\,|M_{\tau_i}| < \infty$, $i = 1, 2$, *and*

$$(2.2) \qquad \liminf_{n \to \infty} \mathbf{E}\,|M_n|\,\mathbf{I}_{\{\tau_i > n\}} = 0, \qquad i = 1, 2.$$

Then $\mathbf{E}\,(M_{\tau_2} \,|\, \mathcal{F}_{\tau_1}) = M_{\tau_1}\,(\geq)$; *in particular,* $\mathbf{E}\,M_{\tau_2} = \mathbf{E}\,M_{\tau_1}\,(\geq)$.

Clearly, (2.2) holds for bounded stopping times $\tau_1 \leq \tau_2 \leq N < \infty$.

§ 3. Local martingales and stochastic integrals

If in Definition 2.1 we get rid of the integrability of $(X_n)_{n \in \mathbb{Z}_+}$, replacing it by the condition

$$\left\{ \mathbf{E}\,(X_n^+ \,|\, \mathcal{F}_{n-1}) < \infty \right\} \cup \left\{ \mathbf{E}\,(X_n^- \,|\, \mathcal{F}_{n-1}) < \infty \right\} = \Omega \quad \text{(a.s.)},$$

then the conditional expectation $\mathbf{E}\,(X_n \,|\, \mathcal{F}_{n-1})$ is well defined as the difference $\mathbf{E}\,(X_n^+ \,|\, \mathcal{F}_{n-1}) - \mathbf{E}\,(X_n^- \,|\, \mathcal{F}_{n-1})$. Consequently, if we require that

$$\mathbf{E}\,(X_n \,|\, \mathcal{F}_{n-1}) = X_{n-1}, \qquad n \in \mathbb{Z}_+,$$

then we arrive at a concept related to a martingale: a *generalized martingale*.

Further, if we apply the *stopping procedure* in Example 2.3 to a stochastic sequence $(X_n)_{n \in \mathbb{Z}_+}$ with $X_0 = 0$ and to a sequence of stopping times increasing to infinity, $\tau_k \uparrow \infty$ ($k \uparrow \infty$), and if it then happens that $X^{\tau_k} = (X_{n \wedge \tau_k}, \mathcal{F}_n)_{n \in \mathbb{Z}_+}$ is a martingale for each $k \geq 1$, then $(X_n)_{n \in \mathbb{Z}_+}$ is called a *local martingale*. Here the sequence $(\tau_k)_{k=1,2,\ldots}$ is said to be *localizing*, and this "procedure for obtaining a local martingale" is called *localization*.

Example 2.1 presented a construction of a martingale with the help of a predictable transformation of a Bernoulli sequence. This leads to the idea of considering the same kind of transformation for arbitrary stochastic sequences. Accordingly, for a predictable sequence $(H_n)_{n \in \mathbb{Z}_+}$ and a stochastic sequence $(Y_n)_{n \in \mathbb{Z}_+}$ we define a new stochastic sequence $((H \cdot Y)_n)_{n \in \mathbb{Z}_+}$ by the formula

$$(2.3) \qquad (H \cdot Y)_n = \sum_{k=0}^{n} H_k\,\Delta Y_k, \qquad \Delta Y_0 = Y_0.$$

The stochastic sequence $(H \cdot Y)$ determined by (2.3) is called the *discrete stochastic integral* of H with respect to Y. The following properties of it are obvious:
  1) it is linear in H and Y;
  2) $\Delta(H \cdot Y)_k = H_k\,\Delta Y_k$, $k \in \mathbb{Z}_+$.

If H is a predictable sequence, and Y is a martingale, then $(H \cdot Y)$ is called a *martingale transformation*. For a square-integrable martingale M and a sequence H such that

$$\mathbf{E}\,\sum_{k=0}^{n} H_k^2\,\Delta\langle M \rangle_k < \infty,$$

the discrete stochastic integral is again a square-integrable martingale, with quadratic characteristic determined by the formula
  3) $\langle H \cdot M \rangle_n = \sum_{k=0}^{n} H_k^2\,\Delta\langle M \rangle_k$.

The following statement holds for the above three different extensions of the class of martingales.

THEOREM 2.3. *The classes of generalized martingales, local martingales, and martingale transformations coincide.*

We remark also that in the case when the set Ω is finite, the class of martingales coincides with the class of local martingales.

§4. Semimartingales

The *variation* of a stochastic sequence X is defined to be the sequence

$$\operatorname{Var}(X)_n = \sum_{k=0}^{n} |\Delta X_k|, \qquad n \in \mathbb{Z}_+.$$

This stochastic sequence is finite for each $n \in \mathbb{Z}_+$, and hence $(\operatorname{Var}(X)_n)_{n \in \mathbb{Z}_+}$ is a "process of bounded variation".

Therefore, the original stochastic sequence has a (nonunique) representation in the form

(2.4) $$X_n = X_0 + A_n + M_n,$$

where A has bounded variation, and M is a local martingale. Starting from the decomposition (2.4), one says that X_n is a *semimartingale*, an essentially broader concept important for random processes with *continuous time*. Next, we rewrite X_n in the form

(2.5) $$X_n = \sum_{k=0}^{n} \Delta X_k \, \mathbf{I}_{\{|\Delta X_k| \le 1\}} + \sum_{k=0}^{n} \Delta X_k \, \mathbf{I}_{\{|\Delta X_k| > 1\}}$$

$$= \sum_{k=0}^{n} \mathbf{E}\big(\Delta X_k \, \mathbf{I}_{\{|\Delta X_k| \le 1\}} \,\big|\, \mathcal{F}_{k-1}\big) + \sum_{k=0}^{n} \Delta X_k \, \mathbf{I}_{\{|\Delta X_k| > 1\}}$$

$$+ \sum_{k=0}^{n} \Big\{ \Delta X_k \, \mathbf{I}_{\{|\Delta X_k| \le 1\}} - \mathbf{E}\big(\Delta X_k \, \mathbf{I}_{\{|\Delta X_k| \le 1\}} \,\big|\, \mathcal{F}_{k-1}\big) \Big\}$$

$$= B_n + \sum_{k=0}^{n} \int_{|x|>1} x \, d\mu_k + \sum_{k=0}^{n} \int_{|x| \le 1} x \, d(\mu_k - \nu_k),$$

where

$$B_n = \sum_{k=0}^{n} \mathbf{E}\,(\Delta X_k \, \mathbf{I}_{\{|\Delta X_k| \le 1\}} \,|\, \mathcal{F}_{k-1}),$$

$$\mu_k(\Gamma) = \mathbf{I}_{\{\Delta X_k \in \Gamma\}}, \qquad \Gamma \in \mathcal{B}(\mathbf{R}^1 \setminus \{0\}),$$

$$\nu_k(\Gamma) = \mathbf{P}\,\{\Delta X_k \in \Gamma \,|\, \mathcal{F}_{k-1}\}.$$

The decomposition (2.5) is unique and is said to be *canonical*.

§5. Change of measure and martingales

An important theorem needed in what follows is *Girsanov's theorem* on transformation of local martingales under a locally absolutely continuous change of measure. As in Example 2.4, we assume that two measures $\mathbf{P}$ and $\mathbf{P}^*$ are given on $(\Omega, \mathcal{F}, \mathbb{F})$ such that $\mathbf{P}^* \overset{\text{loc}}{\ll} \mathbf{P}$, and we let Z_n be the corresponding local density. Define $\alpha_n = Z_n Z_{n-1}^{-1} \mathbf{I}_{\{Z_{n-1} > 0\}}$.

THEOREM 2.4. *Suppose that M $(M_0 = 0)$ is a local martingale with respect to* **P** *such that* $\mathbf{E}\left(\alpha_k|\Delta M_k|\,|\,\mathcal{F}_{k-1}\right) < \infty$ *(a.s.)*, $k \geq 1$. *Then the stochastic sequence*

$$M_n^* = M_n - \sum_{k=0}^{n} \mathbf{E}\left(\alpha_k\,\Delta M_k\,|\,\mathcal{F}_{k-1}\right)$$

is a local martingale with respect to the measure $\mathbf{P}^*$.

PROOF. By Theorem 2.3, it suffices to establish that M_n^* is a generalized martingale. Understanding all the equalities below to be $\mathbf{P}^*$-a.s., we note first of all that for any $\mathcal{F}_n$-measurable random variable Y

$$\mathbf{E}^*(Y\,|\,\mathcal{F}_{n-1}) = Z_{n-1}^{-1}\,\mathbf{E}\left(Y Z_n\,|\,\mathcal{F}_{n-1}\right) = \mathbf{E}\left(Y\alpha_n\,|\,\mathcal{F}_{n-1}\right).$$

Further,

$$\begin{aligned}
\mathbf{E}^*(M_n\,|\,\mathcal{F}_{n-1}) &= \mathbf{E}\left(M_n\alpha_n\,|\,\mathcal{F}_{n-1}\right) \\
&= \mathbf{E}\left(\alpha_n\,\Delta M_n\,|\,\mathcal{F}_{n-1}\right) + \mathbf{E}\left(\alpha_n M_{n-1}\,|\,\mathcal{F}_{n-1}\right) \\
&= \mathbf{E}\left(\alpha_n\,\Delta M_n\,|\,\mathcal{F}_{n-1}\right) + M_{n-1},
\end{aligned}$$

and in view of the condition of the theorem this concludes the proof.

§6. Stochastic equations and stochastic exponentials

Let $U = (U_n)_{n\in\mathbb{Z}_+}$ be a stochastic sequence with $U_0 = 0$. The difference equation

(2.6) $$\Delta X_n = X_{n-1}\,\Delta U_n, \qquad X_0 = 1,$$

is called a discrete (linear) *stochastic differential equation* (with respect to the given sequence U), or (in "integral" form)

$$X_n = 1 + \sum_{k=1}^{n} X_{k-1}\,\Delta U_k, \qquad n \geq 1.$$

The *discrete stochastic exponential* (of U) is defined to be the stochastic sequence

(2.7) $$\mathcal{E}_n(U) = \prod_{k=1}^{n}(1 + \Delta U_k), \qquad \mathcal{E}_0(U) = 1.$$

It can be verified immediately (by induction) that the solution of (2.6) coincides with the exponential (2.7):

$$X_n = \mathcal{E}_n(U), \qquad n \in \mathbb{Z}_+.$$

It also follows from the formula (2.7) that this solution is positive if $\Delta U_k > -1$.

Along with the homogeneous equation (2.6) we consider the nonhomogeneous stochastic equation

(2.8) $$\Delta X_n = \Delta N_n + X_{n-1}\,\Delta U_n, \qquad X_0 = N_0.$$

Its solution is supplied by the formula

(2.9) $$X_n = \mathcal{E}_n(U)\left\{N_0 + \sum_{k=1}^{n} \frac{\Delta N_k}{\mathcal{E}_k(U)}\right\}.$$

In particular, for $N_n \equiv X_0 = N_0$ we have that $X_n = X_0 \, \mathcal{E}_n(U)$.

To prove (2.9) we use mathematical induction. For $n = 1$ we get from (2.9) that

$$\mathcal{E}_1(U) \, N_0 + \mathcal{E}_1(U) \, \mathcal{E}_1^{-1}(U) \, \Delta N_1 = (1 + \Delta U_1)N_0 + \Delta N_1 = N_1 + N_0 \, \Delta U_1.$$

On the other hand, it follows from (2.8) that $\Delta X_1 = \Delta N_1 + X_0 \, \Delta U_1$ and $X_1 = N_1 - N_0 + X_0 + X_0 \, \Delta U_1 = N_1 + N_0 \, \Delta U_1$, and this proves (2.9) for the case $n = 1$.

Next, suppose that (2.9) is true for $n - 1$. Then we get from (2.8) and (2.7) that

$$
\begin{aligned}
X_n &= X_{n-1} + \Delta N_n + X_{n-1} \, \Delta U_n = X_{n-1}(1 + \Delta U_n) + \Delta N_n \\
&= (1 + \Delta U_n)\mathcal{E}_{n-1}(U)\left\{ N_0 + \sum_{k=1}^{n-1} \mathcal{E}_k^{-1}(U) \, \Delta N_k \right\} + \Delta N_n \\
&= \mathcal{E}_n(U)\left\{ N_0 + \sum_{k=1}^{n-1} \mathcal{E}_k^{-1}(U) \, \Delta N_k \right\} + \mathcal{E}_n(U) \, \mathcal{E}_n^{-1}(U) \, \Delta N_n \\
&= \mathcal{E}_n(U)\left\{ N_0 + \sum_{k=1}^{n} \mathcal{E}_k^{-1}(U) \, \Delta N_k \right\},
\end{aligned}
$$

which proves (2.9).

The properties of the stochastic exponential are gathered in the next theorem, where U and V are two given stochastic sequences.

THEOREM 2.5. *The stochastic exponential $\mathcal{E}_n(U)$ has the following properties:*
1) *the multiplication rule for stochastic exponentials is*

$$\mathcal{E}_n(U) \, \mathcal{E}_n(V) = \mathcal{E}_n\big(U + V + [U, V]\big),$$

where $[U, V] = \sum_{k=1}^{n} \Delta U_k \, \Delta V_k$ is the quadratic variation of U and V;
2) *if $\mathcal{E}_n(U) \neq 0$ (or $\Delta U_n \neq -1$), then*

$$\mathcal{E}_n^{-1}(U) = \mathcal{E}_n(-U^*), \quad \text{where} \quad \Delta U_n^* = \Delta U_n - \frac{(\Delta U_n)^2}{1 + \Delta U_n};$$

3) *if $\mathcal{E}_n(U) \neq 0$, then $\mathcal{E}_n(U)$ is a local martingale if and only if U_n is a local martingale;*
4) *if τ_0 is the debut of the set $\{(\omega, n) \colon \mathcal{E}_n(\omega) = 0\}$, then $\mathcal{E}_n(U) = 0$ for all $n \geq \tau_0$.*

PROOF. To prove 1) we rewrite the jump of the product of the exponentials in the form

$$
\begin{aligned}
\Delta\big(\mathcal{E}_n(U) \, \mathcal{E}_n(V)\big) \\
&= \mathcal{E}_{n-1}(U) \, \Delta\mathcal{E}_n(V) + \mathcal{E}_{n-1}(V) \, \Delta\mathcal{E}_n(U) + \Delta\mathcal{E}_n(U) \, \Delta\mathcal{E}_n(V) \\
&= \mathcal{E}_{n-1}(U) \, \mathcal{E}_{n-1}(V)\{\Delta U_n + \Delta V_n + \Delta U_n \, \Delta V_n\}.
\end{aligned}
$$

It is clear from this equality that $\mathcal{E}_n(U) \, \mathcal{E}_n(V)$ satisfies the equation (2.6) with U replaced by $U + V + [U, V]$.

The assertion 2) follows from the multiplication rule:

$$\mathcal{E}_n(U)\,\mathcal{E}_n(-U^*) = \mathcal{E}_n\big(U - U^* - [U, U^*]\big)$$

$$= \mathcal{E}_n\left(U - U + \sum_{k=1}^{n} \frac{(\Delta U_k)^2}{1 + \Delta U_k} - \sum_{k=1}^{n}(\Delta U_k)^2 + \sum_{k=1}^{n} \frac{(\Delta U_k)^3}{1 + \Delta U_k}\right)$$

$$= \mathcal{E}_n(0) = 1.$$

To prove 3) it is convenient to use Theorem 2.3, which establishes that $\mathcal{E}_n(U)$ (or U) is a generalized martingale.

($\Longleftarrow$) From (2.6) we have that

$$\mathbf{E}\Big(\big|\mathcal{E}_n(U)\big| \,\Big|\, \mathcal{F}_{n-1}\Big) \le 1 + \sum_{k=1}^{n} \big|\mathcal{E}_{k-1}(U)\big|\,\mathbf{E}\left(|\Delta U_k| \,\big|\, \mathcal{F}_{k-1}\right) < \infty \quad \text{(a.s.)}.$$

Next, it follows from the same equation that

$$(2.10) \qquad \mathbf{E}\left(\mathcal{E}_n(U) \,\big|\, \mathcal{F}_{n-1}\right) = \mathbf{E}\left(\mathcal{E}_{n-1}(U)\,\Delta U_n \,\big|\, \mathcal{F}_{n-1}\right)$$

$$= \mathcal{E}_{n-1}(U)\,\mathbf{E}\left(\Delta U_n \,\big|\, \mathcal{F}_{n-1}\right) = 0.$$

($\Longrightarrow$) From (2.6), in view of the relation $\mathcal{E}_n(U) \neq 0$, we get that

$$\mathbf{E}\big(|\Delta U_n| \,\big|\, \mathcal{F}_{n-1}\big) \le \big|\mathcal{E}_{n-1}^{-1}(U)\big|\,\mathbf{E}\left(\big|\Delta\mathcal{E}_n(U)\big| \,\Big|\, \mathcal{F}_{n-1}\right) < \infty,$$

and the equality (2.10) concludes the proof.

The property 4) follows immediately from the definition of the stochastic exponential (2.7).

The theorem is proved.

This theorem leads to the following useful fact.

Let us consider a positive stochastic sequence of the form

$$X_n = X_0 + A_n + M_n,$$

where X_0 is a finite ($\mathbf{P}$-a.s.) random variable, A is a predictable stochastic sequence, $A_0 = 0$, M is a local martingale, and $M_0 = 0$.

We use the formula (2.3) to define the new stochastic sequences

$$\Delta\widehat{A}_n = X_{n-1}^{-1}\,\Delta A_n,$$

$$\Delta\widehat{M}_n = (X_{n-1} + \Delta A_n)^{-1}\Delta M_n.$$

Observing that

$$\Delta A_n = X_{n-1}\,\Delta\widehat{A}_n \quad \text{and} \quad X_{n-1}(1 + X_{n-1}^{-1}\,\Delta A_n)\,\widehat{\Delta}M_n = \Delta M_n,$$

we have that

$$\Delta X_n = X_{n-1}\big(\Delta\widehat{A}_n + \Delta\widehat{M}_n + \Delta\,[\widehat{A}, \widehat{M}]_n\big).$$

Consequently,

$$X_n = X_0\,\mathcal{E}_n\big(\widehat{A} + \widehat{M} + [\widehat{A}, \widehat{M}]\big)$$

and, in view of Theorem 2.5,

$$(2.5') \qquad\qquad X_n = X_0\,\mathcal{E}_n(\widehat{A})\,\mathcal{E}_n(\widehat{M}).$$

The decomposition obtained is called the *multiplicative decomposition* of the positive semimartingale X.

§ 7. Itô's formula

The most important technical tool in contemporary stochastic analysis is the change of variables formula, or *Itô's formula*. For an arbitrary $\mathbf{R}^d$-valued stochastic sequence $(X_n)_{n \in \mathbb{Z}_+}$ and a continuously differentiable function $f \colon \mathbf{R}^d \to \mathbf{R}^1$ this formula is the simple identity

$$f(X_n) = f(X_0) + \sum_{k=1}^{n} \sum_{i=1}^{d} \frac{\partial f}{\partial x_i}(X_{k-1}) \, \Delta X_k^i$$

$$+ \sum_{k=1}^{n} \left\{ f(X_k) - f(X_{k-1}) \ - \sum_{i=1}^{d} \frac{\partial f}{\partial x_i}(X_{k-1}) \, \Delta X_k^i \right\}.$$

In particular, we have the following widely used formula for summation (integration) by parts: for two stochastic sequences X and Y,

$$X_n Y_n = X_0 Y_0 + \sum_{k=1}^{n} X_{k-1} \, \Delta Y_k$$

$$+ \sum_{k=1}^{n} Y_{k-1} \, \Delta X_k + [X, Y]_n.$$

§ 8. Optimal stopping of a stochastic sequence

Let $(X_n, \mathcal{F}_n)_{n=0,\dots,N}$ be an integrable stochastic sequence, for which the quantity V^N, called the *value*, is defined by the formula

$$(2.11) \qquad V^N = \sup_{0 \le \tau \le N} \mathbf{E}\, X_\tau,$$

where the supremum is over all stopping times τ. Here, it is clear that the definition of the value is unambiguous, because

$$\mathbf{E}\,|X_\tau| \le \sum_{k=0}^{N} \mathbf{E}\,|X_k| < \infty$$

in view of the integrability of the sequence.

An *optimal* stopping time is defined to be a stopping time τ_N with $0 \le \tau^N \le N$ such that $\mathbf{E}\, X_{\tau^N} = V^N$.

Next, it is convenient to introduce the notation

$$(2.12) \qquad V_n^N = \sup_{n \le \tau \le N} \mathbf{E}\, X_\tau$$

and to denote an optimal time (for this value) by $\tau_n^N \colon \mathbf{E}\, X_{\tau_n^N} = V_n^N$.

It is natural to ask about finding both the values and the optimal stopping times in the problems (2.11)–(2.12). An effective means for this is the following *dynamic programming principle* (the Bellman principle, "backward induction").

First of all, we present some heuristic arguments. If $n = N$, then $V_N^N = \mathbf{E}\, X_N$, and the problem is trivial. Next, let $n = N - 1$; then clearly we must compare

X_{N-1} with the "prediction" of X_N (the conditional expectation $\mathbf{E}\left(X_N \mid \mathcal{F}_{N-1}\right)$).
The following expression is obtained here for the optimal stopping time:

$$
\tau_{N-1}^N = \begin{cases} N-1, & \text{if } X_{N-1} \geq \mathbf{E}\left(X_N \mid \mathcal{F}_{N-1}\right), \\ N, & \text{if } X_{N-1} < \mathbf{E}\left(X_N \mid \mathcal{F}_{N-1}\right). \end{cases}
$$

This idea leads to a formalization of the given principle.

THEOREM 2.6. *Let* $Y_N^N = X_N$, *and let* $Y_n^N = \max\{X_n, \mathbf{E}\left(Y_{n+1}^N \mid \mathcal{F}_n\right)\}$ *for* $n = 0, \ldots, N-1$. *Then the stopping time* $\sigma_n^N = \inf\{k \geq n \colon X_k = Y_k^N\}$ *takes values in* $[n, n+1, \ldots, N]$, *and for any stopping time* $\tau \in [n, N]$

$$
\mathbf{E}\left(X_{\sigma_n^N} \mid \mathcal{F}_n\right) = Y_n^N \geq \mathbf{E}\left(X_\tau \mid \mathcal{F}_n\right).
$$

In particular,

$$
\mathbf{E}\, X_{\sigma_n^N} = \mathbf{E}\, Y_n^N \geq \mathbf{E}\, X_\tau, \quad \text{and} \quad V_n^N = \mathbf{E}\, Y_n^N.
$$

PROOF. The assertion is obvious for $n = N$. Therefore, we assume that it is true for some $1 \leq n \leq N$, and we consider a stopping time τ with $n - 1 \leq \tau \leq N$ and let $\tau' = \tau \vee n$, so that $n \leq \tau' \leq N$. Then for any $A \in \mathcal{F}_{n-1}$

$$
\begin{aligned}
\mathbf{E}\, X_\tau \mathbf{I}_A &= \mathbf{E}\, X_{n-1} \mathbf{I}_{A \cap \{\tau = n-1\}} + \mathbf{E}\, X_{\tau'} \mathbf{I}_{A \cap \{\tau \geq n\}} \\
&= \mathbf{E}\, X_{n-1} \mathbf{I}_{A \cap \{\tau = n-1\}} + \mathbf{E}\, \mathbf{I}_{A \cap \{\tau \geq n\}} \mathbf{E}\Big\{ \mathbf{E}\left(X_{\tau'} \mid \mathcal{F}_n\right) \Big| \mathcal{F}_{n-1} \Big\} \\
&\leq \mathbf{E}\, X_{n-1} \mathbf{I}_{A \cap \{\tau = n-1\}} + \mathbf{E}\, \mathbf{I}_{A \cap \{\tau \geq n\}} \mathbf{E}\left(Y_n^N \mid \mathcal{F}_{n-1}\right) \\
&\leq \mathbf{E}\, Y_{n-1}^N \mathbf{I}_A.
\end{aligned}
$$

Consequently, for any stopping time τ with $n - 1 \leq \tau \leq N$,

$$
\mathbf{E}\left(X_\tau \mid \mathcal{F}_{n-1}\right) \leq Y_{n-1}^N.
$$

Next, we take τ to be the stopping time σ_{n-1}^N. Then $\tau' = \sigma_n^N$ on the set $\{\sigma_{n-1}^N \geq n\}$ and we have that

$$
\begin{aligned}
\mathbf{E}\, X_{\sigma_{n-1}^N} \mathbf{I}_A &= \mathbf{E}\, X_{n-1} \mathbf{I}_{A \cap \{\sigma_{n-1}^N = n-1\}} \\
&\quad + \mathbf{E}\, X_{\sigma_n^N} \mathbf{I}_{A \cap \{\sigma_{n-1}^N \geq n\}} \\
&= \mathbf{E}\, X_{n-1} \mathbf{I}_{A \cap \{X_{n-1} \geq \mathbf{E}\left(Y_n^N (\mathcal{F}_{n-1})\right)\}} \\
&\quad + \mathbf{E}\, \mathbf{E}\left[X_{\sigma_n^N} \mid \mathcal{F}_n\right] \mathbf{I}_{A \cap \{X_{n-1} < \mathbf{E}\left(Y_n^N \mid \mathcal{F}_{n-1}\right)\}} \\
&= \mathbf{E}\, X_{n-1} \mathbf{I}_{A \cap \{X_{n-1} \geq \mathbf{E}\left(Y_n^N \mid \mathcal{F}_{n-1}\right)\}} \\
&\quad + \mathbf{E}\, \mathbf{I}_{A \cap \{X_{n-1} < \mathbf{E}\left(Y_n^N \mid \mathcal{F}_{n-1}\right)\}} \mathbf{E}\left(Y_n^N \mid \mathcal{F}_{n-1}\right) \\
&= \mathbf{E}\, \mathbf{I}_A \max\{X_{n-1}, \mathbf{E}\left(Y_n^N \mid \mathcal{F}_{n-1}\right)\} = \mathbf{E}\, Y_{n-1}^N \mathbf{I}_A.
\end{aligned}
$$

Hence, $\mathbf{E}\left(X_{\sigma_{n-1}^N} \mid \mathcal{F}_{n-1}\right) = Y_{n-1}^N$, which concludes the proof of the theorem.

PROBLEMS

2.1. Prove that $\mathbf{E} M_\tau = \mathbf{E} M_0$ for a martingale $M = (M_n, \mathcal{F}_n)_{0 \le n \le N}$ and for any stopping time $\tau \le N < \infty$.

2.2. Suppose that $\mathbf{P} \overset{\mathrm{loc}}{\sim} \widetilde{\mathbf{P}}$ and $Z_n = d\widetilde{\mathbf{P}}_n / d\mathbf{P}_n$. Prove the following *formula for changing the measure in conditional mathematical expectations*: for a fixed $N \in \mathbb{Z}_+$ and for any integrable $\mathcal{F}_N$-measurable random variable Y

$$Z_{N-1} \widetilde{\mathbf{E}} \left(Y \mid \mathcal{F}_{N-1} \right) = \mathbf{E} \left(Y Z_N \mid \mathcal{F}_{N-1} \right)$$

($\mathbf{P}$ and $\widetilde{\mathbf{P}}$-a.s.).

2.3. Find the relation between two stochastic sequences U_n and $\widehat{U}_n$ such that $\exp\{U_n\} = \mathcal{E}_n(\widehat{U})$, $n \in \mathbb{Z}_+$.

2.4. Suppose that the positive numerical sequence $(\alpha_n)_{0 \le n \le N}$ and the stochastic sequence $(V_n)_{0 \le n \le N}$ correctly determine $A_n = \sum_{k=1}^{n} \mathbf{E}\left(e^{\alpha_k \Delta V_k} - 1 \mid \mathcal{F}_{k-1}\right)$, $n \le N$. Prove that the stochastic sequence $Z_n = \exp\{\sum_{k=1}^{n} \alpha_k \Delta V_k\} \mathcal{E}_n^{-1}(A)$, $Z_0 = 1$, is a martingale with respect to the same filtration $(\mathcal{F}_n)_{n \le N}$. Defining the measure $\mathbf{P}\{A\} = \mathbf{E} Z_N \mathbf{I}_A$ on $\mathcal{F}_N$, prove that (for independent increments ΔV_n) this measure has the form

$$\widetilde{\mathbf{P}}\{A\} = \mathbf{E} \mathbf{I}_A \frac{\exp\{\alpha_N \Delta V_N\}}{\mathbf{E} \exp\{\alpha_N \Delta V_N\}}.$$

We remark that the latter transformation of measures is usually called an *Esscher transform* in *actuarial mathematics*.

In the third lecture the model (3.1) of a discrete financial market is introduced. Such markets will be studied in what follows. The theorems in this lecture on martingale characterization of such key properties of a market as arbitrage and completeness were first presented in [28] and [29]. On this circle of problems see [4], [12], [20]–[24], [37], and [40].

A stochastic model for a financial market. Arbitrage and completeness

§ 1. A model for a market and investment strategies

As a *model for the evolution of prices* (price dynamics) of basic securities on a financial market we consider a system of two discrete stochastic differential equations describing a nonrisky asset B and a risky asset S:

(3.1)
$$\Delta B_n = r_n B_{n-1},$$
$$\Delta S_n = \rho_n S_{n-1}.$$

In addition, it is assumed here that the stochastic base $(\Omega, \mathcal{F}, \mathbb{F}, \mathbf{P})$ is *discrete*: Ω consists of finitely many elements, $|\Omega| < \infty$, and $\mathcal{F} = \mathcal{F}_N$ for some $N \in \mathbb{Z}_+$. For the sums of the first n terms of the (stochastic) sequences $(r_n)_{n \in \mathbb{Z}_+}$ and $(\rho_n)_{n \in \mathbb{Z}_+}$ we introduce the notation

$$U_n = \sum_{k=0}^{n} r_k, \quad V_n = \sum_{k=0}^{n} \rho_k,$$

and we rewrite the solutions of the equations (3.1) according to (2.6), (2.7) in the form of stochastic exponentials:

(3.2)
$$B_n = B_0 \, \mathcal{E}_n(U), \qquad S_n = S_0 \, \mathcal{E}_n(V).$$

We assume that $\mathcal{F}_n = \sigma\{S_0, \ldots, S_n\}$, that is, $\mathcal{F}_n$ is the smallest σ-algebra with respect to which $S_0, \ldots, S_n$ are measurable. A market determined by the equations (3.1) or the formulas (3.2) will be called a (B, S)-*market*.[1]

It might seem that we could get a more general model of a (B, S)-market by assuming only that the prices of the assets B and S are positive. However, in this case we have the *multiplicative representation* (2.5′) for B and S, which again leads to the model (3.1)–(3.2) being considered here.

An *investment strategy* or *portfolio* is defined to be a two-dimensional stochastic sequence $\pi = (\pi_n = (\beta_n, \gamma_n))_{n \in \mathbb{Z}_+}$ whose elements $\beta_n \in \mathcal{F}_n$ and $\gamma_n \in \mathcal{F}_{n-1}$ are interpreted as the amounts of the respective assets B and S at the time $n \in \mathbb{Z}_+$.

The *value* of the portfolio π is defined to be the stochastic sequence $X^\pi = (X_n^\pi)_{n \in \mathbb{Z}_+}$ with $X_n^\pi = \beta_n B_n + \gamma_n S_n$.

The class of portfolios π with the property

$$(3.3) \qquad\qquad B_{n-1}\Delta\beta_n + S_{n-1}\Delta\gamma_n = 0,$$

is said to be *self-financing* and denoted by SF.

We note at once that the value of a self-financing portfolio π admits the representation $X_n^\pi = X_0^\pi + \sum_{k=0}^n (\beta_k \Delta B_k + \gamma_k \Delta S_k)$, which is equivalent to the self-financing condition (3.3), where $\Delta B_0 = \Delta S_0 = 0$.

Among all the portfolios π in SF we single out those that realize an *arbitrage opportunity* of the market in the following sense:

$$X_0^\pi = 0, \quad X_n^\pi \geq 0 \quad \text{for} \quad n \leq N \quad (\mathbf{P}\text{-a.s.})$$

and

$$X_N^\pi > 0 \quad \text{with positive probability.}$$

The economic significance of this definition is already clear from its very statement; namely, an *arbitrage portfolio* provides an opportunity of making a profit without risk.

We denote the class of such portfolios by $\mathrm{SF}_{\mathrm{arb}}$, and we refer to an arbitrage market or a no-arbitrage market depending on whether or not this class is nonempty.

§2. Martingale measures and arbitrage

A probability measure $\mathbf{P}^*$ equivalent to $\mathbf{P}$ is said to be a *martingale* measure, or to be *risk-neutral* if the stochastic sequence $(S_n/B_n)_{n \leq N}$ is a martingale with respect to $\mathbf{P}^*$. The class of these measures is denoted by $\mathbb{P}^*$.

First, we find conditions under which a measure $\mathbf{P}'$ on the market (3.1)–(3.2) is a martingale.

THEOREM 3.1 (criterion for a measure $\mathbf{P}'$ to have the martingale property). *In the model of the market* (3.1) *suppose that the stochastic sequence* $(r_n)_{n \leq N}$ *is predictable and* $r_n > -1$. *Then with respect to* $\mathbf{P}'$,

$$R_n = \frac{S_n}{B_n} \text{ is a martingale} \quad \Longleftrightarrow \quad \left(\sum_{k=0}^n (\rho_k - r_k)\right)_{n \leq N} \text{ is a martingale.}$$

---

[1]To avoid unnecessary technical difficulties we present the following exposition for the case of *one-dimensional* S_n. Here, as will be shown, the most substantial model of a (complete) (B, S)-market is the *binomial model*. However, many of the results carry over adequately to the *vector* case (see, for example, [17], [40], and also Appendix 3).

PROOF. According to the properties of stochastic exponentials, we have from Theorem 2.5 that for all $n \leq N$

$$(3.4) \qquad R_n = \frac{S_n}{B_n} = R_0 \, \mathcal{E}_n(V) \, \mathcal{E}_n^{-1}(U) = \mathcal{E}_n(V) \, \mathcal{E}_n(-U^*)$$

$$= R_0 \, \mathcal{E}_n(V - U^* - [V, U^*])$$

$$= R_0 \, \mathcal{E}_n \left(V - U + \sum_{k=1}^{n} \frac{(\Delta U_k)^2}{1 + \Delta U_k} + [V, -U] \right.$$

$$\left. + \sum_{k=1}^{n} \frac{\Delta V_k \, (\Delta U_k)^2}{1 + \Delta U_k} \right)$$

$$= R_0 \, \mathcal{E}_n \left(\sum_{k=1}^{n} \Delta V_k - \sum_{k=1}^{n} \Delta U_k - \sum_{k=1}^{n} \Delta V_k \, \Delta U_k \right.$$

$$\left. + \sum_{k=1}^{n} \frac{(1 + \Delta V_k)(\Delta U_k)^2}{1 + \Delta U_k} \right)$$

$$= R_0 \, \mathcal{E}_n \left(\sum_{k=1}^{n} \Delta V_k - \sum_{k=1}^{n} \Delta U_k - \sum_{k=1}^{n} \frac{\Delta V_k \, \Delta U_k}{1 + \Delta U_k} \right.$$

$$\left. + \sum_{k=1}^{n} \frac{(\Delta U_k)^2}{1 + \Delta U_k} \right)$$

$$= R_0 \, \mathcal{E}_n \left(\sum_{k=1}^{n} \frac{\Delta V_k - \Delta U_k}{1 + \Delta U_k} \right).$$

It follows from (3.4) and Theorem 2.5 that R_n is a local martingale if and only if $\sum_{k=0}^{n} (\rho_k - r_k)$ is a local martingale. Therefore, the "martingale equivalence" follows from the fact that Ω is finite and $N < \infty$. The theorem is proved.

It is certainly natural to ask about finding a martingale measure $\mathbf{P}^*$ among all the measures equivalent to $\mathbf{P}$. For this let us denote the corresponding local density by $(Z_n)_{n \leq N}$. Then by Theorem 3.1,

(the quotient R is a martingale) $\Longleftrightarrow$ ($V - U$ is a martingale)

with respect to $\mathbf{P}^*$.

For simplicity suppose that V is a martingale already with respect to the original measure $\mathbf{P}$. By Girsanov's theorem, the stochastic sequence

$$V_n^* = V_n - \sum_{k \leq n} \mathbf{E} \, (Z_{k-1}^{-1} Z_k \, \Delta V_k \, | \, \mathcal{F}_{k-1})$$

is a martingale with respect to $\mathbf{P}^*$. Consequently, the measure $\mathbf{P}^*$ must be chosen in such a way that the density satisfies the relation $\Delta U_n = \mathbf{E} \, (Z_{n-1}^{-1} Z_n \, \Delta V_n \, | \, \mathcal{F}_{n-1})$.

The general case is treated similarly.

The next theorem, sometimes called *the fundamental theorem of financial mathematics*, is about the close interconnection between the above concept of arbitrage of a market (a category in economics) and the concept of the martingale property of a measure (a category in mathematics).

THEOREM 3.2. *Suppose that in the market model* (3.1)–(3.2) *the sequence* r_n ($r_n > -1$, $n \le N$) *is deterministic. Then*

$$\mathbb{P}^* \ne \varnothing \iff \mathrm{SF}_{\mathrm{arb}} = \varnothing.$$

PROOF. ($\Longrightarrow$) Let $\mathbf{P}^* \in \mathbb{P}^*$. Then for any strategy $\pi \in \mathrm{SF}$

$$
\begin{aligned}
(3.5) \qquad \Delta X_n^\pi &= \beta_n \Delta B_n + \gamma_n \Delta S_n \\
&= \beta_n r_n B_{n-1} + \gamma_n \rho_n S_{n-1} \\
&= \beta_n r_n B_{n-1} + \gamma_n r_n S_{n-1} + \gamma_n \rho_n S_{n-1} - \gamma_n r_n S_{n-1} \\
&= r_n(\beta_n B_{n-1} + \gamma_n S_{n-1}) + \gamma_n S_{n-1}(\rho_n - r_n) \\
&= r_n X_{n-1}^\pi + \gamma_n S_{n-1}(\rho_n - r_n).
\end{aligned}
$$

Consequently, X_n^π satisfies the nonhomogeneous discrete stochastic differential equation

$$\Delta X_n^\pi = \Delta N_n + X_{n-1}^\pi \Delta U_n, \qquad X_0^\pi = X_0,$$

where $\Delta N_n = \gamma_n S_{n-1}(\rho_n - r_n) = \gamma_n S_{n-1}(\Delta V_n - \Delta U_n)$.

Therefore, using the representation (2.9) for its solution, we get that

$$(3.5') \qquad X_n^\pi = \mathcal{E}_n(U)\left\{ X_0 + \sum_{k=1}^n \mathcal{E}_k^{-1}(U) \Delta N_k \right\}.$$

Since U is deterministic, the martingale property of $\mathbf{P}^*$ and Theorem 3.1 show that

$$(3.6) \qquad \mathbf{E}^* X_n^\pi = \mathcal{E}_n(U)\,\mathbf{E}^* X_0^\pi = 0$$

if $X_0^\pi = 0$.

We assume now that

$$\mathrm{SF}_{\mathrm{arb}} \ne \varnothing \quad \text{and} \quad \pi \in \mathrm{SF}_{\mathrm{arb}}.$$

Then in view of the equivalence $\mathbf{P} \sim \mathbf{P}^*$ we get that $\mathbf{E}^* X_N^\pi > 0$, contradicting (3.6).

The proof of the implication ($\Longleftarrow$) is more difficult and requires a *separation theorem.*

Thus, suppose that $\mathrm{SF}_{\mathrm{arb}} = \varnothing$.

We introduce two nonempty sets of random variables $\xi = \xi(\omega)$ on $(\Omega, \mathcal{F})$:

$$\Sigma_0 = \{\xi \in \mathbf{R}: \text{ there is a } \pi \in \mathrm{SF} \text{ such that } X_0^\pi = 0 \text{ and } X_N^\pi = \xi\},$$
$$\Sigma_1 = \{\xi \ge 0: \mathbf{E}\xi \ge 1\}.$$

We show that

$$(3.7) \qquad \mathrm{SF}_{\mathrm{arb}} = \varnothing \implies \Sigma_0 \cap \Sigma_1 = \varnothing.$$

Suppose that $\Sigma_0 \cap \Sigma_1 \ne \varnothing$. Then by the definitions of Σ_0 and Σ_1, there is a strategy $\pi \in \mathrm{SF}$ such that $X_0^\pi = 0$, $X_N^\pi \ge 0$, and $\mathbf{P}\{X_N^\pi > 0\} > 0$. We establish that this implies the *existence* of a strategy $\overline{\pi}$ in the class $\mathrm{SF}_{\mathrm{arb}}$, which contradicts the assumption that $\mathrm{SF}_{\mathrm{arb}} = \varnothing$.

If $X_n^\pi \ge 0$ for all $0 \le n \le N$, then $\overline{\pi}$ must be taken to be π. Therefore, only the case when $X_n^\pi < 0$ with positive probability at some time n is nontrivial. It is clear that then (since $X_N^\pi \geq 0$ and Ω is finite) there is an $m < N$ such that for some $\omega' \in \Omega$ with $\mathbf{P}\{\omega'\} > 0$

$$X_m^\pi(\omega') = \beta_m(\omega') B_m + \gamma_n(\omega') S_m(\omega') < 0,$$

and $X_n^\pi(\omega) \geq 0$ for $n > m$ and $\omega \in \Omega$.

We define a strategy $\overline{\pi} = (\overline{\pi}_n)_{0 \leq n \leq N}$ with $\overline{\pi}_n = (\overline{\beta}_n, \overline{\gamma}_n)$ as follows. Let $a = X_m^\pi(\omega') \ (< 0)$, $A = \{\omega : X_m^\pi(\omega) = a\}$, and

$$(3.8) \qquad \overline{\beta}_n(\omega) = \mathbf{I}_A(\omega) \left[\beta_n(\omega) - \frac{a}{B_m} \right] \mathbf{I}_{\{n > m\}},$$

$$(3.9) \qquad \overline{\gamma}_n(\omega) = \mathbf{I}_A(\omega) \, \gamma_n(\omega) \, \mathbf{I}_{\{n > m\}}.$$

It is clear that the $\overline{\pi}_n$ are $\mathcal{F}_{n-1}$-measurable. Therefore, to check that $\overline{\pi} \in \mathrm{SF}$ we must (by virtue of the construction (3.8), (3.9) and the assumption that $\pi \in \mathrm{SF}$) show that

$$(3.10) \qquad B_m \, \Delta\overline{\beta}_{m+1} + S_m \, \Delta\overline{\gamma}_{m+1} = 0.$$

This obviously holds on the set $\overline{A}$, while on A

$$\Delta\overline{\beta}_{m+1} = \overline{\beta}_{m+1} = \left[\beta_{m+1} - \frac{a}{B_m} \right],$$

$$B_m \, \Delta\overline{\beta}_{m+1} = B_m \beta_{m+1} - a,$$

and $\Delta\overline{\gamma}_{m+1} = \gamma_{m+1}$. Therefore, by the self-financing property,

$$B_m \, \Delta\overline{\beta}_{m+1} + S_m \, \Delta\overline{\gamma}_{m+1} = B_m \beta_{m+1} + S_m \gamma_{m+1} - a$$
$$= B_m \beta_m + S_m \gamma_m - a = 0$$

on A.

Thus, (3.10) holds, and hence $\overline{\pi} \in \mathrm{SF}$.

We show that $X_n^{\overline{\pi}} \geq 0$ for all $0 \leq n \leq N$. If $\omega \in \overline{A}$, then $X_n^{\overline{\pi}}(\omega) = 0$ for all $0 \leq n \leq N$. But if $\omega \in A$, then for $n \leq m$ it is again true that $X_N^{\overline{\pi}}(\omega) = 0$, while for $n > m$

$$X_n^{\overline{\pi}}(\omega) = \overline{\beta}_n(\omega) \, B_n + \overline{\gamma}_n(\omega) \, S_n$$
$$= \beta_n(\omega) \, B_n + \gamma_n(\omega) \, S_n - \frac{a B_n}{B_m} \geq 0,$$

because $X_n^\pi(\omega) \geq 0$ for $n > m$ and $\omega \in \Omega$, and $a < 0$.

Finally, for $\omega \in A$

$$X_N^{\overline{\pi}}(\omega) = \overline{\beta}_N(\omega) \, B_N + \overline{\gamma}_N(\omega) \, S_N$$
$$= \beta_N(\omega) \, B_N + \gamma_N(\omega) \, S_N - \frac{a B_N}{B_m}$$
$$= X_N^\pi(\omega) - \frac{a B_N}{B_m} > 0.$$

Thus, the implication (3.7) has been established, and hence the assumption of no arbitrage shows that the sets Σ_0 and Σ_1 do not have common elements.

Since Ω is finite, each random variable ξ defined on Ω can be identified with a vector $x = (\xi(\omega_1), \ldots, \xi(\omega_k)) \in \mathbf{R}^k$, where $k = |\Omega|$. Consequently, the sets Σ_0 and Σ_1 can be regarded as disjoint subsets of the finite-dimensional Euclidean space $\mathbf{R}^k$. Here it is clear that Σ_1 is a convex set, while Σ_0 is a linear space. Therefore, according to the separation theorem in a finite-dimensional Euclidean space, there is a linear functional $l = l(x)$, $x \in \mathbf{R}^k$, such that $l(x) = 0$ for $x \in \Sigma_0$ and $l(x) > 0$ for $x \in \Sigma_1$. We remark that the linear functional $l(x)$ on $\mathbf{R}^k$ can be written as the inner product $l(x) = (x, q)$ with some vector $q = (q_1, \ldots, q_k)$, so that $l(x) = (x, q) \equiv \sum_{i=1}^{k} x_i q_i > 0$ for $x = (x_1, \ldots, x_k) \in \Sigma_1$.

If all the $p_i = \mathbf{P}\{\omega_i\}$, $i = 1, \ldots, k$, are positive, then it follows from the definition of Σ_1 that the vectors $(p_1^{-1}, 0, \ldots, 0), (0, p_2^{-1}, 0, \ldots, 0), \ldots, (0, 0, \ldots, p_k^{-1})$ belong to Σ_1, and hence all the q_i, $i = 1, \ldots, k$, are positive, and they can be normalized in such a way that $q_1 + \cdots + q_k = 1$.

We show that the measure $\mathbf{P}^*$ defined on $(\Omega, \mathcal{F})$ by $\mathbf{P}^*\{\omega_i\} = q_i$ is the desired "martingale" measure, that is, we show that the sequence $(S_n/B_n, \mathcal{F}_n, \mathbf{P}^*)$ is a martingale.

Since $(x, q) = 0$ for all $x \in \Sigma_0$, this property means, in terms of the random variables ξ in Σ_0, that $\mathbf{E}^* \xi = 0$. Thus, if π is a self-financing strategy with $X_0^\pi = 0$, then (see the definition of Σ_0) $\mathbf{E}^* X_N^\pi = 0$.

Our subsequent arguments are directed toward using this property to prove that the sequence $(S_n/B_n, \mathcal{F}_n, \mathbf{P}^*)_{0 \leq n \leq N}$ is a martingale. For this it suffices to show (see Example 2.2) that for any stopping time τ (with respect to $\mathbb{F} = (\mathcal{F}_n)_{0 \leq n \leq N}$) such that $0 \leq \tau(\omega) \leq N$ we have the equality

$$(3.11) \qquad\qquad \mathbf{E}^* \left(\frac{S_\tau}{B_\tau} - \frac{S_0}{B_0} \right) = 0.$$

With this goal we take such a stopping time $\widetilde{\tau}$ and construct a strategy $\widetilde{\pi} = (\widetilde{\pi}_n)_{0 \leq n \leq N}$ so that the property $\mathbf{E}^* X_N^{\widetilde{\pi}} = 0$ yields precisely the "martingale" equality (3.11) with $\tau = \widetilde{\tau}$. For example, we can set

$$\widetilde{\beta}_n = \frac{S_{\widetilde{\tau}}}{B_{\widetilde{\tau}}} \mathbf{I}_{\{n > \widetilde{\tau}\}} - \frac{S_0}{B_0}, \qquad \widetilde{\gamma}_n = \mathbf{I}_{\{n \leq \widetilde{\tau}\}}.$$

We remark that, since $\widetilde{\tau}$ is a stopping time, $\widetilde{\beta}_n$ and $\widetilde{\gamma}_n$ are $\mathcal{F}_{n-1}$-measurable. Further, $\widetilde{\pi} = (\widetilde{\pi}_n)$ with $\widetilde{\pi}_n = (\widetilde{\beta}_n, \widetilde{\gamma}_n)$ is a self-financing strategy, because

$$\widetilde{\beta}_n B_n + \widetilde{\gamma}_n S_n = \frac{S_{\widetilde{\tau}}}{B_{\widetilde{\tau}}} \mathbf{I}_{\{n > \widetilde{\tau}\}} B_n + S_n \mathbf{I}_{\{n \leq \widetilde{\tau}\}} - \frac{S_0}{B_0} B_n,$$

$$\widetilde{\beta}_{n+1} B_n + \widetilde{\gamma}_{n+1} S_n = \frac{S_{\widetilde{\tau}}}{B_{\widetilde{\tau}}} \mathbf{I}_{\{n+1 > \widetilde{\tau}\}} B_n + S_n \mathbf{I}_{\{n+1 \leq \widetilde{\tau}\}} - \frac{S_0}{B_0} B_n,$$

and hence

$$B_n \Delta \widetilde{\beta}_{n+1} + S_n \Delta \widetilde{\gamma}_{n+1} = \frac{S_{\widetilde{\tau}}}{B_{\widetilde{\tau}}} \mathbf{I}_{\{\widetilde{\tau} = n\}} B_n - S_n \mathbf{I}_{\{\widetilde{\tau} = n\}} = 0,$$

which means that the strategy $\widetilde{\pi}$ is self-financing.

Finally,

$$
\begin{aligned}
0 = \mathbf{E}^* X_N^{\widetilde{\pi}} &= \mathbf{E}^* \big\{ \widetilde{\beta}_N B_N + \widetilde{\gamma}_N S_N \big\} \\
&= \mathbf{E}^* \bigg\{ \bigg(\frac{S_{\widetilde{\tau}}}{B_{\widetilde{\tau}}} \mathbf{I}_{\{\widetilde{\tau} < N\}} - \frac{S_0}{B_0} \bigg) B_N + \mathbf{I}_{\{\widetilde{\tau} = N\}} S_N \bigg\} \\
&= B_N \, \mathbf{E}^* \bigg\{ \frac{S_{\widetilde{\tau}}}{B_{\widetilde{\tau}}} - \frac{S_0}{B_0} \bigg\} \\
&\quad + \mathbf{E}^* \bigg\{ S_N \mathbf{I}_{\{\widetilde{\tau} = N\}} - \frac{S_{\widetilde{\tau}}}{B_{\widetilde{\tau}}} \mathbf{I}_{\{\widetilde{\tau} = N\}} B_N \bigg\} \\
&= B_N \, \mathbf{E}^* \bigg\{ \frac{S_{\widetilde{\tau}}}{B_{\widetilde{\tau}}} - \frac{S_0}{B_0} \bigg\}.
\end{aligned}
$$

Since $B_N \neq 0$, this proves (3.11) for an arbitrary stopping time $\tau = \widetilde{\tau}$. This proves the implication ($\Longleftarrow$) and with it the theorem.

We remark that the assumption that the predictable sequence $(r_n)_{n \in \mathbb{Z}_+}$ is *non-random* was used only in the proof of the *implication* ($\Rightarrow$) in Theorem 3.2.

§ 3. Martingale measures and completeness

Together with no-arbitrage, another important characterization of a market leading to exact formulas in various financial calculations is its *completeness*.

We say that a (B, S)-market defined on a discrete filtered probability space $(\Omega, \mathcal{F}, \mathbb{F}, \mathbf{P})$ with $|\Omega| < \infty$ is *complete* if for any $\mathcal{F}$-measurable function $f = f(\omega)$ there is a strategy $\pi \in \mathrm{SF}$ whose terminal value reproduces the function f, that is, $X_N^\pi(\omega) = f(\omega)$, $\omega \in \Omega$.

From a nonmathematical point of view the property of completeness ensures the *accessibility* of all assets involved in the market and the *absence of constraints* for investing in these assets.

We have the following theorem on completeness of a (B, S)-market.

THEOREM 3.3. *Suppose that the set $\mathbb{P}^*$ of martingale measures is nonempty, and let $\mathbf{P}^* \in \mathbb{P}^*$. Then the following assertions are equivalent:*
1) *the (B, S)-market is complete;*
2) *the measure $\mathbf{P}^*$ is the only element in $\mathbb{P}^*$;*
3) *every martingale $(M_n, \mathcal{F}_n, \mathbf{P}^*)$, $0 \le n \le N$, admits a representation*

$$
(3.12) \qquad M_n = M_0 + \sum_{k=1}^n \gamma_k(\omega) \, \Delta m_k,
$$

where the random variables $\gamma_k = \gamma_k(\omega)$ are $\mathcal{F}_{k-1}$-measurable, and

$$
\Delta m_k = \frac{S_k}{B_k} - \frac{S_{k-1}}{B_{k-1}} .
$$

We remark that $\Delta m_k = S_{k-1} B_k^{-1}(\rho_k - r_k)$, and in view of Theorem 3.1 the martingale representation (3.12) is equivalent to the representation

$$
(3.12') \qquad M_n = M_0 + \sum_{k=1}^n \widetilde{\gamma}_k \, (\rho_k - r_k),
$$

where $\widetilde{\gamma}_k = \gamma_k S_{k-1} B_k^{-1} \in \mathcal{F}_{k-1}$, which will be used most often in what follows.

PROOF. 1) $\implies$ 2). Assume that, besides $\mathbf{P}^* \in \mathbb{P}^*$, there is another measure $\mathbf{P}^{**} \in \mathbb{P}^*$ such that $\mathbf{P}^{**} \neq \mathbf{P}^*$, that is, there is a set $A \in \mathcal{F}$ for which $\mathbf{P}^{**}\{A\} \neq \mathbf{P}^*\{A\}$.

We take $f(\omega) = \mathbf{I}_A(\omega)B_N$. The assumption of completeness of the market implies that there is a strategy $\pi \in \mathrm{SF}$ such that

$$\mathbf{P}\left\{X_N^\pi(\omega) = \mathbf{I}_A(\omega)B_N\right\} = 1.$$

Since $\mathbf{P}^* \sim \mathbf{P}$ and $\mathbf{P}^{**} \sim \mathbf{P}$, we have that

$$\mathbf{P}^*\{X_N^\pi = \mathbf{I}_A B_N\} = \mathbf{P}^{**}\{X_N^\pi = \mathbf{I}_A B_N\} = 1.$$

By the martingale property of the measures $\mathbf{P}^*$ and $\mathbf{P}^{**}$,

$$\mathbf{E}^*\frac{X_N^\pi}{B_N} = \frac{X_0^\pi}{B_0} \quad \text{and} \quad \mathbf{E}^{**}\frac{X_N^\pi}{B_N} = \frac{X_0^\pi}{B_0}.$$

Consequently, $\mathbf{E}^*\mathbf{I}_A = \mathbf{E}^{**}\mathbf{I}_A$, that is, $\mathbf{P}^*\{A\} = \mathbf{P}^{**}\{A\}$, which contradicts the assumption that $\mathbf{P}^*$ and $\mathbf{P}^{**}$ are different.

2) $\implies$ 1). Let us show that if $\mathbf{P}^*$ is the *unique* martingale measure, then the (B, S)-market is complete.

We define sets of random variables $\xi = \xi(\omega)$ on $(\Omega, \mathcal{F})$:

$$\Sigma_0 = \{\xi \in \mathbf{R}: \text{ there is a } \pi \in \mathrm{SF} \text{ such that } X_0^\pi = 0 \text{ and } X_N^\pi = \xi\},$$
$$\Sigma_2 = \{\xi \in \mathbf{R}: \mathbf{E}^*\xi = 0\}.$$

Obviously, $\Sigma_0 \subseteq \Sigma_2$, and to prove the assertion 2) $\implies$ 1) it suffices to show that 2) $\implies$ "$\Sigma_0 = \Sigma_2$" $\implies$ 1).

We begin with a proof of the second implication.

Let $f = f(\omega)$ be an $\mathcal{F}$-measurable function. Since $\Sigma_0 = \Sigma_2$, the random variable $\xi = f - \mathbf{E}^*f$ is an element of Σ_0. Therefore, there is a strategy $\pi \in \mathrm{SF}$ with $\pi_n = (\gamma_n, \beta_n)_{1 \leq n \leq N}$ such that $X_N^\pi = \xi$. But then the strategy $\widetilde{\pi}$ with $\widetilde{\pi}_n = (\widetilde{\gamma}_n, \widetilde{\beta}_n)_{1 \leq n \leq N}$, where $\widetilde{\gamma}_n = \gamma_n$ and $\widetilde{\beta}_n = \mathbf{E}^*f/B_N + \beta_n$, is self-financing, and $X_N^{\widetilde{\pi}} = f$.

We now show that the coincidence of the sets Σ_0 and Σ_2 follows from the uniqueness of $\mathbf{P}^* \in \mathbb{P}^*$, that is, we show that 2) $\implies \Sigma_0 = \Sigma_2$. With this goal we identify, as in the proof of Theorem 3.2, each random variable ξ defined on Ω with the vector $x = (\xi(\omega_1), \ldots, \xi(\omega_k)) \in \mathbf{R}^k$, where $k = |\Omega|$, and we identify the probability measure $\mathbf{P}^*$ on $(\Omega, \mathcal{F})$ with the vector $q^* = (q_1^*, \ldots, q_k^*) \in \mathbf{R}^k$, where $q_i^* = \mathbf{P}^*\{\omega_i\} > 0$, $i \leq k$. It is clear that Σ_0 and Σ_2 are linear subspaces of $\mathbf{R}^k$.

If $\Sigma_0 \neq \Sigma_2$, then since $\Sigma_0 \subseteq \Sigma_2$, there is a nonzero vector $\widetilde{x} \in \Sigma_2$ orthogonal to the set Σ_0:

$$(\widetilde{x}, x) = \sum_{i=1}^k \widetilde{x}_i x_i = 0, \qquad x \in \Sigma_0.$$

Let $\varepsilon > 0$ be such that $\widetilde{q}_i = q_i^* - \varepsilon\widetilde{x}_i > 0$ for all $i \leq k$.

Let $\widetilde{q} = (\widetilde{q}_1, \ldots, \widetilde{q}_k)$. In this case if $x \in \Sigma_0$, then $(\widetilde{q}, x) = (q^*, x) = 0$.

As in the proof of Theorem 3.2, we can show that the measure $\widetilde{\mathbf{P}}$ on $(\Omega, \mathcal{F})$ defined by $\widetilde{\mathbf{P}}\{\omega_i\} = \delta\widetilde{q}_i$, where $\delta = (\widetilde{q}_1 + \cdots + \widetilde{q}_k)^{-1}$, is a martingale measure, that is, the sequence $(S_n/B_n, \mathcal{F}_n, \widetilde{\mathbf{P}})$ is a martingale.

By virtue of the assumed uniqueness of $\mathbf{P}^*$ we get the equality $\mathbf{P}^* = \widetilde{\mathbf{P}}$, which is equivalent to the condition that $q^* = \delta\widetilde{q} = \delta q^* - \varepsilon\,\delta\,\widetilde{x}$, or

(3.13)
$$(1 - \delta)\,q^* = \varepsilon\,\delta\,\widetilde{x}.$$

However, since $\widetilde{x} \in \Sigma_2$, the vectors q^* and $\widetilde{x}$ are orthogonal, and hence the equality (3.13) is possible only when $\delta = 1$ and the vector $\widetilde{x}$ is *zero*.

The contradiction obtained shows that $\Sigma_0 = \Sigma_2$.

1) $\Longrightarrow$ 3). We show that the property of completeness of the market ensures that every martingale $M = (M_n, \mathcal{F}_n, \mathbf{P}^*)$, $0 \leq n \leq N$, can be represented in the form (3.12).

Let M be a martingale and let $f = M_N B_N$. Since the market is complete, there is a strategy $\pi \in \mathrm{SF}$ whose value X_N^π at the time N is exactly equal to f: $X_N^\pi(\omega) = f(\omega)$.

The martingale property of the sequence $(X_n^\pi/B_n, \mathcal{F}_n, \mathbf{P}^*)$ implies the equality $X_n^\pi/B_n = M_n$. Consequently,

$$
\begin{aligned}
M_{n+1} - M_n &= \frac{X_{n+1}^\pi}{B_{n+1}} - \frac{X_n^\pi}{B_n} \\
&= \frac{\beta_{n+1}\,B_{n+1} + \gamma_{n+1}\,S_{n+1}}{B_{n+1}} - \frac{\beta_{n+1}\,B_n + \gamma_{n+1}\,S_n}{B_n} \\
&= \gamma_{n+1}\left(\frac{S_{n+1}}{B_{n+1}} - \frac{S_n}{B_n}\right),
\end{aligned}
$$

which leads to the "integral" representation (3.12).

3) $\Longrightarrow$ 1). We show that the possibility of representing every martingale in the form (3.12) implies the completeness of the (B, S)-market.

Let $f = f(\omega)$ be a random variable on $(\Omega, \mathcal{F})$. We define a martingale $M = (M_n, \mathcal{F}_n, \mathbf{P}^*)$, $0 \leq n \leq N$, by setting

$$M_n = \mathbf{E}^*\left[\frac{f}{B_N}\,\Big|\,\mathcal{F}_n\right].$$

In view of our assumption,

$$M_n = M_0 + \sum_{k=1}^{n}\gamma_k\left(\frac{S_k}{B_k} - \frac{S_{k-1}}{B_{k-1}}\right).$$

We determine a portfolio π^* with $\pi_n^* = (\beta_n^*, \gamma_n^*)$ such that $\gamma_n^* = \gamma_n$ and $\beta_n^* = M_n - \gamma_n S_n/B_n$ for $n \leq N$.

Then $\pi^* \in \mathrm{SF}$, because for $n \leq N$

$$
\begin{aligned}
S_{n-1}\,\Delta\gamma_n^* &+ B_{n-1}\,\Delta\beta_n \\
&= S_{n-1}\,\Delta\gamma_n + B_{n-1}\left(\Delta M_n - \Delta\left(\gamma_n\frac{S_n}{B_n}\right)\right) \\
&= S_{n-1}\,\Delta\gamma_n + B_{n-1}\left[\gamma_n\,\Delta\frac{S_n}{B_n} - \Delta\left(\gamma_n\frac{S_n}{B_n}\right)\right] = 0, \\
X_n^{\pi^*} &= \beta_n^*\,B_n + \gamma_n^*\,S_n = M_n B_n.
\end{aligned}
$$

In particular, $X_N^{\pi^*} = M_N B_N = f$, which concludes the proof of the implication 3) $\Longrightarrow$ 1), and with it the whole theorem.

Problems

3.1. Find a general formula connecting the predictable sequence (r_n), the stochastic sequence (ρ_n) in (3.1)–(3.2), and the local density Z_n of the martingale measure $\mathbf{P}^*$ with respect to the original measure $\mathbf{P}$.

3.2. Suppose that a (B,S)-market is determined by a d-dimensional sequence $(S_n)_{n \leq N}$ of prices and $B_n \equiv 1$. Prove that in general a martingale measure exists in neither of the two "infinite" markets with $d = \infty$ and with $N = \infty$.

3.3. Suppose that a stochastic base $(\Omega, \mathcal{F}, (\mathcal{F}_n)_{n \in \mathbb{Z}_+}, \mathbf{P})$ is given. For each $N = 1, 2, \dots$ let $\mathcal{F}_n^N = \mathcal{F}_n$ for $n \leq N$, and let $\mathbf{P}^N = \mathbf{P} \,|\, \mathcal{F}_N^N$. On each base $(\Omega, \mathcal{F}_N^N, (\mathcal{F}_n^N)_{n \leq N}, \mathbf{P}^N)$ let us consider a no-arbitrage (B^N, S^N)-market, where $B_n^N \equiv 1$, S_n^N is $\mathcal{F}_n^N$-measurable, and $\mathbf{P}^{*N}$ is a measure with respect to which S^N is a martingale. Then the portfolio π^N is determined by the predictable sequence $(\gamma_n^N)_{n \leq N}$, and its value is equal to

$$X_n^N = X_0^N + \sum_{k=1}^{n} \gamma_k^N \, \Delta S_k^N, \qquad n \leq N.$$

Prove that:

a) X_n^N and $X_n^N Z_n^N$ are martingales with respect to $\mathbf{P}^{*N}$ and $\mathbf{P}^N$, where $Z_n^N = d\mathbf{P}_n^{*N}/d\mathbf{P}^N$;

b) $(Z_N, \mathcal{F}_N, \mathbf{P})_{N \geq 1} = (Z_N^N, \mathcal{F}_N^N, \mathbf{P})_{N \geq 1}$ is a martingale, and $Z_\infty = \lim_{N \to \infty} Z_N$, $0 \leq \mathbf{E} Z_N \leq 1$;

c) the condition $\mathbf{P}\{Z_\infty > 0\} = 1$ is sufficient for the sequence (π^N) of strategies to be an *asymptotically arbitrage* sequence in the sense that $X_0^N \to 0$ ($\mathbf{P}$-a.s.) as $N \to \infty$, $X_N^N \geq -C_N$, $C_N \downarrow 0$ ($N \to \infty$), and for some $\varepsilon > 0$

$$\limsup_{N \to \infty} \mathbf{P}^N \{ X_N^N \geq \varepsilon \} > 0.$$

The fourth lecture is an investigation of the pricing of European options, regarded on a complete no-arbitrage market. General formulas are given for computing a fair price and hedging strategies. In the particular case when this market is binomial, the given results lead to the well-known Cox–Ross–Rubinstein formula ([12], [18], [19], [21], [23], [31]). The problem of pricing options is considered together with hedging "in probability" ([5], [6]), and an example is given of pricing a "currency option" [26].

Pricing European options in complete markets. The binomial model and the Cox–Ross–Rubinstein formula

§ 1. Contingent claims and European options

We consider a (finite) financial market (3.1)–(3.2). A *contingent claim* with repayment date N is defined to be a pair (f, N), where f is an $\mathcal{F}_N$-measurable nonnegative random variable. The participant on the market who must repay the given claim has to organize his investment activities so that the corresponding investment portfolio π has value $X_N^\pi \geq f$. The procedure of building a portfolio for which the contingent claim is attainable is called *hedging* the claim and the portfolio itself is called a *hedging portfolio*.

The nature of contingent claims can be fairly arbitrary. One of the most important problems connected with the hedging of contingent claims arises in connection with the *pricing of options*. On a (B, S)-market we assume an issuer has issued a derivative security (for purchase, sale, and so on) of some asset. In order to become the *holder* of such a security it is necessary first to pay to the issuer a certain *premium* $\mathbb{C}$. Here the *right* is acquired to exercise the given security at the time N and to receive a payment in the amount of f. Such a derivative security is called a *European option* (on the purchase, sale, and so on) of an asset, and the transaction itself is called a *contract with an option*.

It is clear that most important here is the question of *the price* at which to sell and buy an option and *how to hedge* the contingent claim with respect to the given option. First, it is necessary to give exact mathematical definitions of these objects.

DEFINITION 4.1. Suppose that on a (B, S)-market (3.1)–(3.2) an initial value of capital $x > 0$ is given along with a contingent claim (f, N). A self-financing portfolio π is called an $(x, f, N)-hedge$ if the value X^π satisfies the conditions that for all $\omega \in \Omega$

(4.1)
$$X_0^\pi = x,$$
$$X_N^\pi \geq f(\omega).$$

A hedge is said to be *minimal* if equality is attained in (4.1). In this case one refers to the *attainability* of the contingent claim (f, N).

Let $\Pi(x, f, N)$ stand for the set of all (x, f, N)-hedges.

DEFINITION 4.2. The *investment cost* of a contingent claim (f, N) is defined to be the quantity

(4.2)
$$\mathbb{C}(N) = \inf \left\{ x > 0 \colon \Pi(x, f, N) \neq \varnothing \right\}.$$

Note that since Ω is finite, $\mathbb{C}(N)$ is bounded.

Suppose now that (f, N) is a contingent claim on an option (on the purchase, sale, and so on, of some asset). It is clear that the formula (4.2) realizes the idea of satisfying both the seller (he can "attain" the claim (f, N) on the given market) and the buyer (in a certain sense he pays a minimal premium to the seller). For this reason the quantity $\mathbb{C}(N)$ is called the *fair price of the option*.

EXAMPLE 4.1(Options to buy and to sell).
An *option to buy* (a call option) is an option with contingent claim $f = (S_N - K)^+$ according to which its holder has the opportunity to acquire shares (or another asset) at a fixed price K (the strike price of the asset) at a time N (the exercise time), at which time the market price of the shares is equal to S_N. It is clear that this option is exercised if $S_N > K$ and is not exercised if $S_N \leq K$.
An *option to sell* (a put option) is an option according to which the holder has the right to sell shares at a fixed price K, and hence the corresponding contingent claim has the form $f = (K - S_N)^+$.

EXAMPLE 4.2(Exotic options).
A *collar option*: $f = \min \left\{ \max (K_1, S_N), K_2 \right\}$;
A *Boston option*: $f = \max \left\{ S_N - K_1, 0 \right\} - (K_2 - K_1)$, where $K_1 < K_2$ are positive constants.

In Examples 4.1 and 4.2 the contingent claim is entirely determined by the price of the asset at the exercise time of the option. However, the contingent claim can depend in an essential way on the whole "path" $S_0, S_1, \ldots, S_N$ of prices or on part of the path.

EXAMPLE 4.3 (look-back options, or path-dependent options).
A *look-back option to buy*:

$$f = (S_N - K_N)^+, \quad \text{where} \quad K_N = \min \left\{ S_0, \ldots, S_N \right\}.$$

A *look-back option to sell*:

$$f = (K_N - S_N)^+, \quad \text{where} \quad K_N = \max \left\{ S_0, \ldots, S_N \right\}.$$

An *Asian* arithmetic *option to buy and sell*, respectively:

$$f = (\overline{S}_N - K)^+, \quad f = (K - \overline{S}_N)^+,$$

where $\overline{S}_N = (1/(N+1)) \sum_{k=0}^{N} S_k$.

§2. General formulas for computing prices and hedging strategies for European options

We assume the *no-arbitrage* and *completeness* conditions in a (B, S)-market (3.1)–(3.2), with r_n a *deterministic* sequence such that $r_n > -1$ for all $n = 0, 1, \ldots, N$. Consequently, by Theorems 3.2 and 3.3, there is a unique martingale measure $\mathbf{P}^*$, and the *martingale representation* (3.12) is valid. Under these assumptions we have the following theorem, which gives general formulas for the fair price of a European option, a minimal hedge, and the corresponding value.

THEOREM 4.1. *Suppose that the (B, S)-market (3.1)–(3.2) does not admit arbitrage opportunities and is complete, and let r_n be a deterministic sequence $r_n > -1$ for $n = 0, 1, \ldots, N$. Then for a European option with contingent claim (f, N):*

1) the fair price is given by the formula

$$(4.3) \qquad \mathbb{C}(N) = \mathbf{E}^* \mathcal{E}_N^{-1}(U) f = \mathcal{E}_N^{-1}(U) \mathbf{E}^* f;$$

2) there is a minimal $(\mathbb{C}(N), f, N)$-hedge $\pi^* = (\pi_n^*)_{n \leq N} = (\beta_n^*, \gamma_n^*)_{n \leq N}$ whose value X^{π^*} can be represented in the form

$$(4.4) \qquad X_n^{\pi^*} = \mathbf{E}^* (\mathcal{E}_N^{-1}(U) \mathcal{E}_n(U) f \mid \mathcal{F}_n),$$

and the components of π^* satisfy the relations

$$(4.5) \qquad \gamma_n^* = \frac{\widetilde{\gamma}_n B_n}{S_{n-1}} \in \mathcal{F}_{n-1}, \quad \beta_n^* = \frac{X_{n-1}^{\pi^*} - \gamma_n^* S_{n-1}}{B_{n-1}} \in \mathcal{F}_{n-1},$$

where $(\widetilde{\gamma}_n)_{n \leq N}$ is the predictable sequence in the decomposition (3.12′) for the martingale $(\mathbf{E}^*(B_N^{-1} f \mid \mathcal{F}_n))_{n=0,\ldots,N}$.

PROOF. Let $\pi \in \Pi(x, f, N)$. We consider the *discounted* (that is, equalized in price at different moments of time) value $M_n^\pi = X_n^\pi / B_n$ of this hedge. By (3.5′),

$$(4.6) \qquad M_n^\pi = X_0 B_0^{-1} + \sum_{k=1}^{n} B_k^{-1} \gamma_k S_{k-1} (\rho_k - r_k).$$

Since $\mathbf{P}^*$ is a martingale measure, it follows from Theorem 3.1 that the stochastic sequence $(\sum_{k=0}^{n} (\rho_k - r_k))_{n \leq N}$ is a martingale (with respect to $\mathbf{P}^*$). Further, using the finiteness of Ω and the predictability of $\gamma_n S_{n-1}$, we get from (4.6) that $(M_n^\pi)_{n \leq N}$ is a martingale with respect to $\mathbf{P}^*$. Consequently, $\mathbf{E}^* M_N^\pi = X_0 B_0^{-1}$, or

$$(4.7) \qquad x = X_0 = \mathbf{E}^* \mathcal{E}_N^{-1}(U) X_N^\pi = \mathcal{E}_N^{-1}(U) \mathbf{E}^* X_N^\pi.$$

It follows from (4.7) and (4.1) that for the (x, f, N)-hedge π

$$(4.8) \qquad x \geq \mathcal{E}_N^{-1}(U) \mathbf{E}^* f,$$

and

$$(4.9) \qquad x = \mathcal{E}_N^{-1}(U) \mathbf{E}^* f$$

in the case when this hedge is minimal.

Using (4.2), we get from (4.8) that

$$(4.10) \qquad \mathbb{C}(N) \geq \mathcal{E}_N^{-1}(U)\, \mathbf{E}^* f.$$

We now prove the existence of a minimal (x, f, N)-hedge π^* if the condition (4.9) holds.

Let us consider the martingale

$$(4.11) \qquad M_n^* = \mathbf{E}^*\big(B_N^{-1} f \mid \mathcal{F}_n\big), \qquad n = 0, 1, \ldots, N.$$

In view of (4.9) it is clear that

$$(4.12) \qquad M_0^* = \mathbf{E}^* B_N^{-1} f = B_N^{-1} \mathbf{E}^* f = B_0^{-1} \mathcal{E}_N^{-1} \mathbf{E}^* f = \frac{x}{B_0}, \qquad M_N^* = B_N^{-1} f.$$

We now introduce the notation

$$\gamma_k^* = \widetilde{\gamma}_k\, B_k S_{k-1}^{-1}, \qquad k = 1, \ldots, N,$$

and rewrite the representation $(3.12')$ for the martingale M^* in the form

$$M_n^* = \frac{x}{B_0} + \sum_{k=1}^{n} \frac{\gamma_k^*\, S_{k-1}}{B_k}\, \Delta(\rho_k - r_k).$$

We show now that there is a self-financing strategy π^* such that the corresponding discounted value coincides with M^*:

$$(4.13) \qquad \frac{X_n^{\pi^*}}{B_n} = M_n^* \quad \text{for all} \quad n = 0, \ldots, N.$$

Indeed, having $\gamma_1^* = \gamma_1 B_1 / S_0$, we set

$$\beta_1^* = \frac{x - \gamma_1^* S_0}{B_0}, \quad \pi_1^* = (\beta_1^*, \gamma_1^*),$$

and, correspondingly,

$$X_1^{\pi^*} = \beta_1^* B_1 + \gamma_1^* S_1.$$

Then for the discounted value we have

$$\begin{aligned} M_1^{\pi^*} &= \frac{X_1^{\pi^*}}{B_1} = \beta_1^* + \gamma_1^* \frac{S_1}{B_1} = \frac{x}{B_0} + \gamma_1^*\left(\frac{S_1}{B_1} - \frac{S_0}{B_0}\right) \\ &= \frac{x}{B_0} + \frac{\widetilde{\gamma}_1 B_1}{S_0}\left(\frac{S_1}{B_1} - \frac{S_0}{B_0}\right) = \frac{x}{B_0} + \widetilde{\gamma}_1\left(\frac{S_1}{S_0} - \frac{B_1}{B_0}\right) \\ &= \frac{x}{B_0} + \widetilde{\gamma}_1\,(\rho_1 - r_1) = M_1^*. \end{aligned}$$

Continuing this process with β_n^* and γ_n^* chosen according to the formulas (4.5), we see by induction that the equality (4.13) holds.

Consequently, for the hedge constructed we have by (4.11) that

$$\frac{X_n^{\pi^*}}{B_n} = M_n^{\pi^*} = M_n^* = \mathbf{E}^*\big(B_N^{-1} f \mid \mathcal{F}_n\big).$$

Using the exponential expression (3.2) for B_n, we get the formula (4.4) for the value of this hedge:

$$X_n^{\pi^*} = \mathbf{E}^*\big(\mathcal{E}_N^{-1}(U)\, \mathcal{E}_N(U) f \mid \mathcal{F}_n\big).$$

In particular, for the initial and final value we have

$$X_0^{\pi^*} = \mathcal{E}_N^{-1}(U)\,\mathbf{E}^*\,(f\mid\mathcal{F}_n) = x,$$
$$X_N^{\pi^*} = f,$$

which shows the minimality of the hedge constructed and, together with the relation (4.10), leads to the equality (4.3).

The theorem is proved.

We note that the purely theoretical formulas (4.3) and (4.4) also give a natural basis for the *approximate* computation of the fair price and of the value of a hedging portfolio (for example, with the help of the *Monte Carlo method*).

§3. The binomial model of a (B, S)-market. Its properties of no-arbitrage and completeness

We consider a (B, S)-market (3.1) in the case when the sequence $(r_n)_{n\leq N}$ is nonrandom and identically equal to a positive constant r (*the interest rate of a bank account B*), while the other sequence $(\rho_n)_{n\leq N}$ is assumed to be random and to take only two values a and b, where

$$(4.14) \qquad\qquad -1 < a < r < b.$$

We rewrite ρ_n in the form

$$(4.15) \qquad\qquad \rho_n = \frac{a+b}{2} + \frac{b-a}{2}\,\varepsilon_n$$

and

$$(4.16) \qquad\qquad \rho_n = a + (b-a)\delta_n,$$

where

$$\varepsilon_n = \begin{cases} +1 \\ -1 \end{cases} \Longleftrightarrow \delta_n = \begin{cases} 1 \\ 0 \end{cases} \Longleftrightarrow \rho_n = \begin{cases} b \\ a \end{cases}.$$

The assumptions and concepts (4.14)–(4.16) lead to the following refinement of the stochastic base on which the model is being considered. Namely, in our case it is natural to take Ω to be the set $\{+1, -1\}^N$, that is, the set of sequences $(\varepsilon_1, \ldots, \varepsilon_N)$, and to take the probability measures $\mathbf{P}$ such that with respect to them $(\rho_n)_{n\leq N}$ is a sequence of independent identically distributed random variables taking only the two values a and b:

$$(4.17) \qquad 0 < q = \mathbf{P}\,\{\rho_n = a\}, \quad 0 < p = \mathbf{P}\,\{\rho_n = b\}, \qquad p + q = 1.$$

If we now define $\mathcal{F}$ to be the collection of all subsets of Ω, and the filtration $\mathbb{F} = (\mathcal{F}_n)_{n\leq N}$ to be the filtration generated by these random variables ($\mathcal{F}_0 = (\varnothing, \Omega)$, $\mathcal{F}_n = \sigma(\varepsilon_1, \ldots, \varepsilon_n)$), then we arrive at a model of the market called the *binomial* model.

It will be shown below how the general computational formulas for this commonly used model of a financial market can be refined. But first, we prove the following theorem.

THEOREM 4.2. *The binomial market* (3.1)–(3.2) *with parameters a, b, r satisfying the condition* (4.14) *is a no-arbitrage and complete market.*

PROOF. We note first that any probability measure $\mathbf{P}$ on the sequence space Ω introduced above is completely determined by the probability p in (4.17). We take p to be the value $p^* = (r - a)/(b - a)$, and we denote the corresponding measure by $\mathbf{P}^*$. Then

$$\mathbf{E}^* (\rho_1 - r) = a (1 - p^*) + b p^* - r = (b - a) p^* - (r - a) = 0.$$

Consequently, the sequence $(\sum_{k=1}^{n}(\rho_k - r))_{n \leq N}$ is a martingale with respect to $\mathbf{P}^*$. By Theorem 3.1, $\mathbf{P}^*$ is a martingale measure, and by Theorem 3.2, the market does not admit arbitrage opportunities.

The completeness of the binomial market is a consequence of Theorem 3.3 and the following *martingale representation lemma*.

LEMMA 4.1. *Let* $(\Omega, \mathcal{F}, \mathbf{P})$ *be a probability space, let* $\rho = (\rho_n)_{n \leq N}$ *be a sequence of independent identically distributed random variables satisfying* (4.14) *and* (4.17), *and let* $p = (r - a)/(b - a)$. *Let* $\mathbb{F} = (\mathcal{F}_n)_{n \leq N}$ *and* $\mathcal{F}_n = \sigma\{\rho_1, \ldots, \rho_n\}$. *Then an arbitrary martingale* $M = (M_n)_{n \leq N}$ *with* $\mathbf{E} M_0 = 0$ *has a representation* (*in the form of a discrete stochastic integral*)

$$(4.18) \qquad M_n = \sum_{k=1}^{n} \alpha_k \, \Delta m_k,$$

where $(\alpha_n)_{n \leq N}$ *is a predictable sequence, and* $m = (m_n)_{n \leq N} = (\sum_{k=1}^{n}(\rho_k - r))_{n \leq N}$ *is a "basic" martingale.*

PROOF. The M_n are $\mathcal{F}_n$-measurable, so there are functions $f_n = f_n(x_1, \ldots, x_n)$ with x_i equal to either a or b such that $M_n(\omega) = f_n(\rho_1(\omega), \ldots, \rho_n(\omega))$ for all $\omega \in \Omega$.

Further, if the representation (4.18) were true, then

$$(4.19) \qquad \Delta M_n(\omega) = \alpha_n(\omega) \, \Delta m_n,$$

which is equivalent to the relations

$$f_n(\rho_1(\omega), \ldots, \rho_{n-1}(\omega), b) - f_{n-1}(\rho_1(\omega), \ldots, \rho_{n-1}(\omega)) = \alpha_n(\omega)(b - r),$$
$$f_n(\rho_1(\omega), \ldots, \rho_{n-1}(\omega), a) - f_{n-1}(\rho_1(\omega), \ldots, \rho_{n-1}(\omega)) = \alpha_n(\omega)(a - r),$$

and hence

$$(4.20)$$
$$\alpha_n(\omega) = \frac{f_n(\rho_1(\omega), \ldots, \rho_{n-1}(\omega), b) - f_{n-1}(\rho_1(\omega), \ldots, \rho_{n-1}(\omega))}{b - r}$$
$$= \frac{f_n(\rho_1(\omega), \ldots, \rho_{n-1}(\omega), a) - f_{n-1}(\rho_1(\omega), \ldots, \rho_{n-1}(\omega))}{a - r}.$$

Using these heuristic arguments, we establish (4.20). Indeed, from the martingale property we have that ($\mathbf{P}$-a.s.)

$$\mathbf{E}\left(f_n(\rho_1(\omega), \ldots, \rho_n(\omega)) - f_{n-1}(\rho_1(\omega), \ldots, \rho_{n-1}(\omega)) \,\Big|\, \mathcal{F}_{n-1} \right) = 0,$$

or ($\mathbf{P}$-a.s.)

$$p \, f_n(\rho_1(\omega), \ldots, \rho_{n-1}(\omega), b) + (1 - p) \, f_n(\rho_1(\omega), \ldots, \rho_{n-1}(\omega), a)$$
$$= f_{n-1}(\rho_1(\omega), \ldots, \rho_{n-1}(\omega)).$$

The last relation can be rewritten in the form

$$\frac{f_n(\rho_1(\omega),\ldots,\rho_{n-1}(\omega),b) - f_{n-1}(\rho_1(\omega),\ldots,\rho_{n-1}(\omega))}{1-p}$$
$$= \frac{f_{n-1}(\rho_1(\omega),\ldots,\rho_{n-1}(\omega)) - f_n(\rho_1(\omega),\ldots,\rho_{n-1}(\omega),a)}{p},$$

and in view of the choice $p = (r - a)/(b - a)$ we arrive at (4.20). Going now in the "backward direction" from (4.20), we get (4.19), and hence also (4.18).

The lemma is proved, and with it Theorem 4.2.

Since $\mathcal{F}_n = \sigma(\rho_1,\ldots,\rho_n) = \sigma(\varepsilon_1,\ldots,\varepsilon_n) = \sigma(\delta_1,\ldots,\delta_n)$, we can use the relations (4.15) and (4.16) to rewrite the representation (4.18) in the form

$$(4.21) \qquad M_n = \sum_{k=1}^{n} \alpha_k^{(\rho)}\,\Delta m_k^{(\rho)} = \sum_{k=1}^{n} \alpha_k^{(\varepsilon)}\,\Delta m_k^{(\varepsilon)} = \sum_{k=1}^{n} \alpha_k^{(\delta)}\,\Delta m_k^{(\delta)},$$

where $\alpha_k^{(\rho)} = \alpha_k^{(\rho)}(\rho_1,\ldots,\rho_{k-1})$, $\alpha_k^{(\varepsilon)} = \alpha_k^{(\varepsilon)}(\varepsilon_1,\ldots,\varepsilon_{k-1})$, $\alpha_k^{(\delta)} = \alpha_k^{(\delta)}(\delta_1,\ldots,\delta_{k-1})$, $\Delta m_k^{(\rho)} = \rho_k - r$, $\Delta m_k^{(\varepsilon)} = \varepsilon_k - (2p - 1)$, and $\Delta m_k^{(\delta)} = \delta_k - p$.

It turns out that the coefficients in the decomposition (4.21) admit the following simple description "in binomial terms".

LEMMA 4.2. *Suppose that M_N has the structure*

$$(4.22) \qquad M_N = g\left(\Delta_N(\omega)\right),$$

where $\Delta_N = \delta_1 + \cdots + \delta_N$, and g is some function. Then for all $k \leq N$ $(\Delta_0 = 0)$

$$(4.23) \qquad \alpha_k^{(\delta)} = G_{N-k}(\Delta_{k-1}; p),$$

where

$$(4.24) \qquad G_n(x; p) = \sum_{k=0}^{n} \left[g\left(x + k + 1\right) - g\left(x + k\right)\right]\binom{n}{k}p^k(1-p)^{n-k}.$$

PROOF. It follows from the equality $\Delta M_n = \alpha_n^{(\delta)}\Delta m_n^{(\delta)}$, the representation $M_n = \mathbf{E}\left(M_N\,|\,\mathcal{F}_n\right)$, and the condition (4.22) that

$$(4.25)$$
$$\alpha_n^{(\delta)} = \frac{\mathbf{E}\left(M_N\,|\,\delta_1,\ldots,\delta_{n-1},1\right) - \mathbf{E}\left(M_N\,|\,\delta_1,\ldots,\delta_{n-1}\right)}{1-p}$$
$$= \frac{\mathbf{E}\left(g\left(\Delta_N\right)\,|\,\delta_1,\ldots,\delta_{n-1},1\right) - \mathbf{E}\left(g\left(\Delta_N\right)\,|\,\delta_1,\ldots,\delta_{n-1}\right)}{1-p}.$$

Further, on the set $\{\omega\colon \Delta_{n-1} = x, \delta_n = 1\}$

$$\mathbf{E}\left(g\left(\Delta_N\right)\,\big|\,\mathcal{F}_n\right) = \mathbf{E}\,g(x + 1 + \Delta_N - \Delta_n),$$
$$\mathbf{E}\left(g\left(\Delta_N\right)\,\big|\,\mathcal{F}_{n-1}\right) = \mathbf{E}\,g(x + \Delta_N - \Delta_{n-1}) = p\,\mathbf{E}\,g(x + 1 + \Delta_N - \Delta_n)$$
$$+ (1-p)\,\mathbf{E}\,g(x + \Delta_N - \Delta_n).$$

Consequently,

$$\mathbf{E}\big(g\,(\Delta_N)\,\big|\,\mathcal{F}_n\big) - \mathbf{E}\big(g\,(\Delta_N)\,\big|\,\mathcal{F}_{n-1}\big)$$
$$= (1-p)\,\mathbf{E}\,\big[g\,(x+1+\Delta_N - \Delta_n) - g\,(x+\Delta_N - \Delta_n)\big]$$
$$= (1-p)\sum_{k=0}^{N-n}\big[g\,(x+1+k) - g\,(x+k)\big]\binom{N-n}{k}p^k(1-p)^{N-n-k}$$
$$= (1-p)\,G_{N-n}(x;\,p)$$

and in view of (4.25) we get (4.23).

The lemma is proved.

§4. Options with contingent claims of the form $f = f(S_N)$ in the binomial model

The fair price and the value of a hedging portfolio admit further specification for options with contingent claim $f = f(S_N)$, considered in the framework of the *binomial model* (3.1) and (4.14), (4.15). For example, the formulas (4.3) and (4.4), which hold in this case because of Theorems 4.2 and 4.1, can be transformed to

$$\mathbb{C}\,(N) = (1+r)^{-N}\mathbf{E}^* f(S_N), \quad X_n^{\pi^*} = (1+r)^{-(N-n)}\mathbf{E}^*\big(f(S_N)\,\big|\,\mathcal{F}_n\big).$$

Next, we introduce the functions

$$(4.26)\qquad F_n(x;\,p) = \sum_{k=0}^{n} f\Big(x(1+b)^k(1+a)^{n-k}\Big)\binom{n}{k}p^k(1-p)^{n-k}.$$

Note that

$$\prod_{n<k\leq N}(1+\rho_k) = (1+b)^{\Delta_N - \Delta_n}(1+a)^{(N-n)-(\Delta_N - \Delta_n)}.$$

Consequently $(p^* = (r-a)/(b-a))$,

$$\mathbf{E}^* f\Big(x\prod_{n<k\leq N}(1+\rho_k)\Big) = F_{N-n}(x;\,p^*)$$

and

$$(4.27)\qquad X_n^{\pi^*} = (1+r)^{-(N-n)}\mathbf{E}^*\big(f(S_N)\,\big|\,\mathcal{F}_n\big)$$
$$= (1+r)^{-(N-n)}\mathbf{E}^*\Big(f\Big(S_n\prod_{n<k\leq N}(1+\rho_k)\Big)\,\Big|\,\rho_1,\ldots,\rho_n\Big)$$
$$= (1+r)^{-(N-n)}F_{N-n}(S_n;\,p^*).$$

In particular,

$$(4.28)\qquad\qquad \mathbb{C}\,(N) = X_0^{\pi^*} = (1+r)^{-N}F_N(S_0;\,p^*).$$

It remains to go into the structure of the minimal hedge $\pi^* = (\pi_n^*) = (\beta_n^*, \gamma_n^*)$ in Theorem 4.1.

We consider the discounted value of this hedge at the exercise time N of the option,

$$M_N = \frac{X_N^{\pi^*}}{B_N} = \frac{f(S_N)}{B_N} = \frac{f\left(S_0(1+b)^{\Delta_N}(1+a)^{N-\Delta_N}\right)}{B_N} = g(\Delta_N),$$

and we get by (4.13) and (4.21) that

$$M_N = M_0 + \sum_{n=1}^{N} \frac{\gamma_n^* S_{n-1}}{B_n}(\rho_n - r) = M_0 + \sum_{n=1}^{N} \frac{\gamma_n^* S_{n-1}}{B_n}(b-a)(\delta_n - p^*).$$

It follows from Lemma 4.2 and the representation (4.21) that

(4.29) $$\gamma_n^* = \frac{\alpha_n^{(\delta)} B_n}{S_{n-1}(b-a)} = \frac{G_{N-n}(\Delta_{n-1}; p^*)B_n}{S_{n-1}(b-a)},$$

where $G_n(x; p)$ is defined according to the formula (4.24) with the function $g(x) = B_N^{-1} f(S_0(1+b)^x(1+a)^{N-x})$.

It remains to express (4.29) in terms of the "initial" function F_{N-n}. To do this we observe that

$$G_{N-n}(x; p) = B_N^{-1} \sum_{k=0}^{N-n} \left[f\left(S_0(1+a)^N \left(\frac{1+b}{1+a}\right)^{x+k+1}\right) \right.$$

$$\left. - f\left(S_0(1+a)^N \left(\frac{1+b}{1+a}\right)^{x+k}\right) \right] \binom{N-n}{k} p^k (1-p)^{N-n-k}.$$

From this, by the equality

$$S_{n-1} = S_0(1+a)^{n-1}\left(\frac{1+b}{1+a}\right)^{\Delta_{n-1}},$$

we get for $x = \Delta_{n-1}$ that

$$G_{N-n}(\Delta_{n-1}; p) = B_N^{-1} \sum_{k=0}^{N-n} \left[f\left(S_{n-1}(1+a)^{N-(n-1)}\left(\frac{1+b}{1+a}\right)^{k+1}\right) \right.$$

$$\left. - f\left(S_{n-1}(1+a)^{N-(n-1)}\left(\frac{1+b}{1+a}\right)^{k}\right) \right] \binom{N-n}{k} p^k (1-p)^{N-n-k}$$

$$= B_N^{-1} \sum_{k=0}^{N-n} \left[f\left(S_{n-1}(1+a)^{N-n}\left(\frac{1+b}{1+a}\right)^{k}(1+b)\right) \right.$$

$$\left. - f\left(S_{n-1}(1+a)^{N-n}\left(\frac{1+b}{1+a}\right)^{k}(1+a)\right) \right] \binom{N-n}{k} p^k (1-p)^{N-n-k}$$

$$= \frac{F_{N-n}(S_{n-1}(1+b); p) - F_{N-n}(S_{n-1}(1+a); p)}{B_N}.$$

Consequently,

(4.30) $$\gamma_n^* = \gamma_n^*(S_{n-1}) = \frac{G_{N-n}(\Delta_{n-1}; p^*)B_n}{S_{n-1}(b-a)} = (1+r)^{-(N-n)}$$

$$\times \frac{F_{N-n}(S_{n-1}(1+b); p^*) - F_{N-n}(S_{n-1}(1+a); p^*)}{S_{n-1}(b-a)}.$$

We now find the first component of the hedging portfolio from (4.27) and (4.30):

(4.31)

$$
\beta_n^* = \frac{X_{n-1}^{\pi^*} - \gamma_n^* S_{n-1}}{B_{n-1}} = \frac{F_{N-n+1}(S_{n-1}; p^*)}{B_N}
$$

$$
- (1+r)^{-(N-n)} \frac{F_{N-n}(S_{n-1}(1+b); p^*) - F_{N-n}(S_{n-1}(1+a); p^*)}{B_{n-1}(b-a)}
$$

$$
= B_N^{-1} \Big\{ F_{N-n+1}(S_{n-1}; p^*) - (1+r)\Big[F_{N-n}\big(S_{n-1}(1+b); p^*\big)
$$

$$
- F_{N-n}\big(S_{n-1}(1+a); p^*\big)\Big] \Big\}.
$$

As a result, if we combine the formulas (4.27) and (4.28), (4.30) and (4.31), we get the following theorem on the price structure of a minimal hedging portfolio and its value in the case under consideration.

THEOREM 4.3. *Suppose that on the binomial* (B, S)*-market* (3.1)–(3.2) *and* (4.14), (4.15) *a European option with contingent claim* $(f(S_N), N)$ *is considered. Then*

1) *the fair price of the option is determined by the formula*

$$
\mathbb{C}(N) = (1+r)^{-N} F_N(S_0; p^*),
$$

where $p^* = (r-a)/(b-a)$, *and* $F_N(x; p)$ *is defined in* (4.26);

2) *the value of a minimal hedge* π^* *is described by the formulas*

$$
X_n^{\pi^*} = (1+r)^{-(N-n)} F_{N-n}(S_n; p^*);
$$

3) *a minimal hedge* $\pi^* = (\pi_n^*)_{n \le N} = (\beta_n^*, \gamma_n^*)_{n \le N}$ *exists, and its components are given by the formulas* (4.30) *and* (4.31).

§5. The Cox–Ross–Rubinstein formula

We consider a European option to buy with the contingent claim $(f, N) = ((S_N - K)^+, N)$. In this case

$$
F_N(S_0; p^*) = \sum_{k=0}^{N} \binom{N}{k} (p^*)^k (1-p^*)^{N-k} \max\left\{ 0, S_0(1+a)^N \left(\frac{1+b}{1+a}\right)^k - K \right\}.
$$

Let

$$
k_0 = \min\left\{ k \in \mathbb{Z}_+ : S_0(1+a)^N \left(\frac{1+b}{1+a}\right)^k > K \right\}.
$$

It is clear that $F_N(S_0; p^*) = 0$ for $k_0 > N$, and hence $\mathbb{C}(N) = 0$.

Let $k_0 \le N$. Then

(4.32) $\mathbb{C}(n) = (1+r)^{-N} F_N(S_0; p^*)$

$$
= S_0 \sum_{k=k_0}^{N} \binom{N}{k} (p^*)^k (1-p^*)^{N-k} \left(\frac{1+a}{1+r}\right)^N \left(\frac{1+b}{1+a}\right)^k
$$

$$
- K(1+r)^{-N} \sum_{k=k_0}^{N} \binom{N}{k} (p^*)^k (1-p^*)^{N-k}.
$$

Setting

$$(4.33) \qquad \widetilde{p} = \frac{1+b}{1+r} p^*, \quad \mathbb{B}(j, N; p) = \sum_{k=j}^{N} \binom{N}{k} p^k (1-p)^{N-k},$$

we get from (4.32) the following theorem on the fair price of a European option to buy.

THEOREM 4.4 (Cox–Ross–Rubinstein formula). *For a European option to buy with contingent claim $F(S_N) = (S_N - K)^+$ the fair price is determined by the formula*

$$(4.34) \qquad \mathbb{C}(N) = S_0 \, \mathbb{B}(k_0, N; \widetilde{p}) - K(1+r)^{-N} \mathbb{B}(k_0, N; p^*),$$

where $k_0 = 1 + \big[\ln \frac{K}{S_0(1+a)^N} \big/ \ln \frac{1+b}{1+a} \big]$, and $\mathbb{C}(N) = 0$ if $k_0 > N$.

It is interesting to note that the formula (4.34) directly implies the formula for the fair price $\mathbb{P}(N)$ of a European option to sell with the contingent claim $f(S_N) = (K - S_N)^+$.

This is clear from the following equality, which connects the prices $\mathbb{P}(N)$ and $\mathbb{C}(N)$ and is called *"call-put parity"*: since $\max\{0, K - S_N\} = \max\{S_N - K, 0\} - S_N + K$, we have that

$$\begin{aligned} \mathbb{P}(N) &= \mathbf{E}^*(1+r)^{-N} \max\{0, K - S_N\} \\ &= \mathbb{C}(N) - \mathbf{E}^*(1+r)^{-N} S_N + K(1+r)^{-N} \\ &= \mathbb{C}(N) - S_0 + K(1+r)^{-N}. \end{aligned}$$

§ 6. An example of option pricing on a currency market

Let S_n be the cost of \$100 in Swiss francs SFR, where $S_0 = 150\,\text{SFR}$, and at the next time $n = 1$ this quantity can become either $180\,\text{SFR}$ or $90\,\text{SFR}$, which corresponds to either an increase or a decrease in the exchange rate of the dollar with respect to the Swiss franc. Further,

$$S_1 = S_0(1 + \rho_1),$$

where ρ_1 is equal to either $b = 1/5$ or $a = -2/5$. Suppose that the interest rate r is equal to 0, and the deposit in the bank account is constant and equal to $B_0 = 1$.

An option to buy (dollars) is offered with exercise date $N = 1$ and exercise price $K = 150\,\text{SFR}$, that is, $f = (S_1 - 150)^+\,\text{SFR}$.

According to the theory above (see (4.34)), to determine the fair price $\mathbb{C}(N)$ we define

$$p^* = \frac{r - a}{b - a} = \frac{2/5}{3/5} = 2/3$$

and the quantity $k_0 = k_0(-2/5, 1/5, 150 = S_0, 150 = K) = 1$. Therefore, the formula (4.34) gives the value

$$\mathbb{C}(1) = S_0 \, p^*(1 + b) - K p^* = S_0 \, p^* b = 150 \cdot \frac{2}{3} \cdot \frac{1}{5} = 20.$$

Accordingly, in "fair" trading of the indicated option the seller has initial capital $X_0 = 20\,\text{SFR}$, and hence β_0 and γ_0 can be chosen so that $\beta_0^* = 0$ and $\gamma_0^* = 2/15$.

As for the next value (β_1^*, γ_1^*) of the minimal hedge, we have by (4.30) that

$$
\begin{aligned}
\gamma_1^* &= \frac{F_0(S_0\,(1+b); p^*) - F_0(S_0\,(1+a); p^*)}{S_0\,(b-a)} \\
&= \frac{f(S_0\,(1+b)) - f(S_0\,(1+a))}{S_0\,(b-a)} = \frac{f(S_0\,(1+b))}{S_0\,(b-a)} \\
&= \frac{\max\{0, S_0\,(1+b) - K\}}{S_0\,(b-a)} = \frac{b}{b-a} = 1/3.
\end{aligned}
$$

Since $X_0 = \beta_1^* B_0 + \gamma_1^* S_0$ and $B_0 = 1$, we have that $\beta_1^* = X_0 - \gamma_1^* S_0 = 20 - \frac{1}{3} \cdot 150 = -30$.

The latter means that to attain his contingent claim the seller of the option, along with getting the premium of 20 SFR, borrows 30 SFR more. The total capital of 50 SFR "is invested" in dollars at the exchange rate of 150 SFR= \$100, that is, \$33.33 is acquired.

When a new exchange rate is announced for the dollar at the time $N = 1$, two situations are possible.

1) The exchange rate becomes \$100 = 180 SFR, and in this case the option is certainly exercised. The value of the contingent claim is equal to 30 SFR, and the value of the (optimal) portfolio chosen by the issuer is equal to

$$
X_1^* = \beta_1^* B_1 + \gamma_1^* S_1 = -30 \cdot 1 + \frac{1}{3} \cdot 180 = 30 \text{ SFR},
$$

so the contingent claim is safely repaid.

2) The exchange rate is \$100 = 90 SFR, and in this case the contingent claim f is equal to 0. Therefore, the issuer must only return a debt of 30 SFR, which he can do by changing \$33.33 at this exchange rate and getting exactly 30 SFR.

§7. Pricing and hedging contingent claims attainable with positive probability

In the course of this lecture the problem of pricing and hedging has been solved for contingent claims "attainable with probability 1". The approach presented below, which is based on statistical analogies, assumes that a contingent claim is attainable "in probability". It turns out that a satisfactory solution of the problem posed is possible for less initial capital than in the case considered above.

Suppose that on a complete (B, S)-market (3.1)–(3.2) with deterministic $r_n > -1$ and $B_0 = 1$ we are given a contingent claim (f, N) such that $\mathbf{E}^* f > 0$, where $\mathbf{E}^*$, as earlier, is the average with respect to the martingale measure $\mathbf{P}^*$. With the given contingent claim we associate the following class of self-financing strategies:

$$
\mathrm{SF}(f, N) = \{\pi \in \mathrm{SF}\colon X_N^\pi \geq f - \mathbf{E}^* f\}.
$$

We set some *level of significance* $\alpha \in (0, 1)$ of the contingent claim (f, N) and call a strategy $\pi \in \mathrm{SF}(f, N)$ an α-(x, f, N)-*hedge*, or simply an α-*hedge* if for $\mathbf{P}' = \mathbf{P}$ and $\mathbf{P}' = \mathbf{P}^*$

(4.35) $$\mathbf{P}'\{X_N^\pi \geq f\} \geq 1 - \alpha.$$

The set of α-hedges is denoted by $\Pi(x, f, N, \alpha)$, and the α-*price* of the contingent claim (f, N) is defined to be

$$
\mathbb{C}\,(N, \alpha) = \inf\,\{x > 0\colon \Pi(x, f, N, \alpha) \neq \varnothing\}.
$$

First of all, we establish that for any portfolio $\pi \in \mathrm{SF}(f, N)$

$$(4.36) \qquad \mathbf{P}^*\{X_N^\pi \geq f\} \leq \frac{X_0^\pi}{\mathbb{C}},$$

where under the formulated conditions $\mathbb{C} = \mathbf{E}^* \mathcal{E}_N^{-1}(U) f$ is the fair price of a European option with claim (f, N).

Indeed, by Chebyshev's inequality, the definition of the class $\mathrm{SF}(f, N)$, and the martingale property of the measure $\mathbf{P}^*$, the formula $(3.5')$ gives us (4.36):

$$\mathbf{P}^*\{X_N^\pi \geq f\} = \mathbf{P}^*\{X_N^\pi - f + \mathbf{E}^* f \geq \mathbf{E}^* f\}$$
$$\leq \frac{\mathbf{E}^*(X_N^\pi - f + \mathbf{E}^* f)}{\mathbf{E}^* f} = \frac{\mathbf{E}^* X_N^\pi}{\mathbf{E}^* f}$$
$$= \frac{\mathbf{E}^* \mathcal{E}_N^{-1}(U) \, X_N^\pi}{\mathbf{E}^* \mathcal{E}_N^{-1}(U) \, f} = \frac{X_0^\pi}{\mathbb{C}} \, .$$

A necessary condition for a strategy π in the class $\mathrm{SF}(f, N)$ to be an α-(x, f, N)-hedge follows from (4.35) and (4.36):

$$(1 - \alpha) \leq \mathbf{P}^*\{X_N^\pi \geq f\} \leq \frac{x}{\mathbb{C}},$$

or

$$(4.37) \qquad x \geq (1 - \alpha)\, \mathbb{C}.$$

As in the proof of Theorem 4.1, an α-hedge π_α with equality in (4.37) will be constructed in this case.

For the original measure $\mathbf{P}$ and the martingale measure $\mathbf{P}^*$ we define the density

$$Z_N = \frac{d\mathbf{P}}{d\mathbf{P}^*} \, .$$

It is clear that there exists a $\lambda = \lambda(\alpha)$ such that

$$(4.38) \qquad \mathbf{P}^*\{Z_N \geq \lambda\} = 1 - \alpha,$$

and with the aim of simplifying the arguments we assume that $\lambda \geq 1$.

Let us consider the martingales

$$M_n^\alpha = \mathbf{E}^* \left(\mathbf{I}_{\{Z_N > \lambda(\alpha)\}} \mid \mathcal{F}_n \right) \quad \text{and} \quad M_n^\mathbb{C} = \mathbf{E}^* \left(\mathcal{E}_N^{-1}(U) f \mid \mathcal{F}_n \right),$$

which, since the market is complete, admit representations

$$M_n^\alpha = \alpha + \sum_{k=1}^{n} \varphi_k \, \mathcal{E}_k^{-1}(U) \, S_{k-1} \, (\rho_k - r_k),$$
$$M_n^\mathbb{C} = \mathbb{C} + \sum_{k=1}^{n} \gamma_k^* \, \mathcal{E}_k^{-1}(U) \, S_{k-1} \, (\rho_k - r_k)$$

with predictable sequences (φ_k) and (γ_k^*).

For the initial value $X_0 = (1 - \alpha)\mathbb{C}$, we define the portfolio $\pi_\alpha = (\beta_n^\alpha, \gamma_n^\alpha)_{n \leq N}$ by the formulas

$$(4.39) \qquad \gamma_n^\alpha = \gamma_n^* - \varphi_n \mathbb{C}, \quad \beta_n^\alpha = \frac{X_{n-1}^{\pi_\alpha} - \gamma_n^\alpha S_{n-1}}{\mathcal{E}_{n-1}(U)} \, .$$

We show that the so-called portfolio π_α is an α-$((1 - \alpha)\mathbb{C}, f, N)$-hedge.

Indeed, it follows from (3.5′) in view of (4.39) that

$$(4.40) \qquad \mathcal{E}_n^{-1}(U)\, X_n^{\pi_\alpha} = (1-\alpha)\mathbb{C} + \sum_{k=1}^{n} \mathcal{E}_k^{-1}(U)\, \gamma_k^\alpha\, S_{k-1}\,(\rho_k - r_k)$$

$$= (1-\alpha)\mathbb{C} + \sum_{k=1}^{n} \mathcal{E}_k^{-1}(U)\, S_{k-1}\, \gamma_k^*\,(\rho_k - r_k)$$

$$- \sum_{k=1}^{n} \mathcal{E}_k^{-1}(U)\, S_{k-1}\, \varphi_k\, \mathbb{C}\,(\rho_k - r_k)$$

$$= (1-\alpha)\mathbb{C} - (1-\alpha)\,\mathbb{C} + M_n^{\mathbb{C}} - \mathbb{C}\,\mathbf{M}_n^\alpha.$$

From (4.40),

$$\mathcal{E}_N^{-1}(U)\, X_N^{\pi_\alpha} = \mathcal{E}_N^{-1}(U)\, f - \mathbb{C}\,\mathbf{I}_{\{Z_N < \lambda\}}$$

and hence

$$\mathcal{E}_N^{-1}(U)\, X_N^{\pi_\alpha} \ge \mathcal{E}_N^{-1}(U)\, f - \mathbb{C}.$$

The last inequality means that $\pi_\alpha \in \mathrm{SF}(f, N)$.

Further, it follows from (4.38) that

$$(4.41) \qquad \mathbf{P}^*\{X_N^{\pi_\alpha} \ge f\} = \mathbf{P}^*\{f - \mathbb{C}\,\mathcal{E}_N(U)\,\mathbf{I}_{\{Z_N < \lambda\}} \ge f\}$$

$$= \mathbf{P}^*\{\mathbf{I}_{\{Z_N < \lambda\}} \le 0\} = \mathbf{P}^*\{Z_N \ge \lambda\} = (1-\alpha).$$

Finally, we get from (4.38) and (4.41) that

$$(4.42) \qquad \mathbf{P}\{X_N^{\pi_\alpha} \ge f\} = \mathbf{E}^*\,\mathbf{I}_{\{X_N^{\pi_\alpha} \ge f\}} Z_N$$

$$\ge \mathbf{E}^*\,\mathbf{I}_{\{X_N^{\pi_\alpha} \ge f\}} \mathbf{I}_{\{Z_N \ge \lambda\}} Z_N$$

$$\ge \lambda\,(1-\alpha) \ge 1 - \alpha.$$

The relations (4.41) and (4.42) show that the condition (4.35) holds for the strategy π_α, and hence π_α is an α-$((1-\alpha)\mathbb{C}, f, N)$-hedge.

What has been obtained shows that it is possible to hedge a contingent claim *with a specified probability* $(1-\alpha)$. Further, *the initial funds can be reduced by the amount* $\alpha\,\mathbb{C}$, though with a *risk* α the accepted contingent claim cannot be repaid.

PROBLEMS

4.1. Prove that on a no-arbitrage (B, S)-market we have for a standard European option to buy (sell) that $\mathbb{C}(N_2) \ge \mathbb{C}(N_1)$ (respectively, $\mathbb{P}(N_2) \ge \mathbb{P}(N_1)$) when $N_2 \ge N_1$.

4.2. Prove that the fair price $\mathbb{C} = \mathbb{C}(N, S_0, K)$ of a standard European option to buy, where N is the exercise time, S_0 is the initial price of a share, and K is the exercise price, has the following properties:

  a) $\mathbb{C}(S_0, K)$ is monotone in S_0 and K;

  d) $\mathbb{C}(S_0, K)$ is convex in S_0 and K;

  c) $\mathbb{C}(\lambda S_0, \lambda K) = \lambda\,\mathbb{C}(S_0, K)$ for $\lambda > 0$.

4.3. Prove that a minimal $(\mathbb{C}(N), f, N)$-hedge is unique.

4.4 Suppose that on a financial market with a bank account $\Delta B_n = rB_{n-1}$ two exchange rates are given: (for example, rubles/dollars (USA) or rubles/DM):

$$\Delta S_N^1 = S_{n-1}^1 \rho_n^1, \quad \Delta S_N^2 = S_{n-1}^2 \rho_n^2,$$

where (ρ^i) $(i = 1, 2)$ are sequences of Bernoulli random variables. Obtain a formula for the exchange rate of dollars (USA)/DM under the assumption that $\operatorname{cov}(\rho_n^1, \rho_n^2) = C_n$.

The fifth lecture presents a theory of pricing and hedging American options under conditions of a complete market (3.1). A special case is a "Russian option" [3]. This problem is shown to be connected in a natural way with the *optimal stopping* problem. The exposition of the material is based on [9] and [12].

Pricing and hedging American options in complete markets

§ 1. Dynamic contingent claims and American options

The contingent claims (f, N) considered up to this point have had a static character in the sense that the exact exercise date N and the amount of the payments $f \in \mathcal{F}_N$ were fixed in advance. We can substantially broaden the class of contingent claims if we assume that $f = (f_n)_{n=0,1,\ldots,N}$ is a nonnegative *stochastic sequence* of length N. Such contingent claims (f, N) will be called *dynamic contingent claims* with *maturity date N*.

Connected in a natural way with dynamic contingent claims (f, N) are the derivative securities for which the holder has the right to exercise them at *any time* $n = 0, 1, \ldots, N$ and to receive a payment in the amount of f_n. And the seller of such a security, who receives a premium C at its sale (the cost of the security) is obligated to handle it in such a way that the value X_n^π of his portfolio at any time n exceeds f_n, that is, he must ensure *hedging* of the contingent claim. Such securities are called *American options*.

Thus, American options present their holders with greater freedom in choosing the *exercise time τ*. Further, the holder of the option makes his decision on the basis of the available information up to that time, and hence the *exercise time* is a *stopping time*.

We consider a finite (B, S)-market (3.1)–(3.2).

DEFINITION 5.1. Suppose that $x > 0$ and $f = (f_n)_{n \leq N}$ is a dynamic contingent claim. A strategy $\pi = (\pi_n)_{n \leq N} \in \mathrm{SF}$ is called an *American (x, f, N)-hedge* if for any $\omega \in \Omega$

$$X_0^\pi(\omega) = x$$

and

$$X_n^\pi(\omega) \geq f_n(\omega) \quad \text{for all} \quad n \leq N.$$

In this case we have the inequality $X_\tau^\pi(\omega) \geq f_\tau(\omega)$ for any stopping time $\tau \leq N$.

As shown earlier, the procedure of constructing a hedge is called *hedging* the (dynamic) contingent claim (f, N).

The set of American (x, f, N)-hedges will be denoted by $\Pi(x, f, N)$.

A hedge is said to be *minimal* if there is a stopping time τ such that

$$X_\tau^\pi(\omega) = f_\tau(\omega) \quad \text{for all} \quad \omega \in \Omega.$$

In precise analogy with the "static" case of European options we give the following definition.

DEFINITION 5.2. Consider an American option with (dynamic) contingent claim (f, N). The quantity

$$\mathbb{C}(N) = \inf \{x > 0 \colon \Pi(x, f, N) \neq \varnothing\}$$

is called the *fair price* or the *rational* cost of the option with contingent claim (f, N).

EXAMPLE 5.1. An American *option to buy* is defined to be a dynamic contingent claim

$$f = (f_n)_{n \leq N}, \quad \text{where} \quad f_n = (S_n - K)^+,$$

while an *option to sell* is defined to be the claim $f_n = (K - S_n)^+$.

EXAMPLE 5.2. A *Russian option* is defined to be an American option with a dynamic contingent claim of the form

$$f_n = \sup_{m \leq n} S_m.$$

It is often convenient to regard American options as "perpetual", that is, $f = (f_n)_{n \in \mathbb{Z}_+}$ is an infinite stochastic sequence.

§2. Pricing American options as an optimal stopping problem

Suppose that we are given a (complete and finite) (B, S)-market determined by (3.1)–(3.2), on which we consider an American option with (dynamic) contingent claim (f, N). If the holder of the option exercises it at the stopping time τ, then the seller, according to the contract with this option, must pay the sum of f_τ.

To price a given option directly or reduce it to a known mathematical problem, we denote the unique martingale measure on the market (3.1)–(3.2) by $\mathbf{P}^*$. Further, assuming that the sequence $r_n > -1$, $n \leq N$, is deterministic, we have for the portfolio $\pi \in \mathrm{SF}$ that its discounted value

$$M^\pi = (M_n^\pi = X_n^\pi / B_n)_{n \leq N}$$

is a martingale with respect to $\mathbf{P}^*$. Consequently, by Theorem 2.2 (see also (3.5′)),

$$\mathbf{E}^* M_\tau^\pi = M_0^\pi,$$

or

(5.1) $$X_0^\pi = \mathbf{E}^* \mathcal{E}_\tau^{-1}(U) X_\tau^\pi.$$

For $\pi \in \Pi(x, f, N)$ in (5.1) it follows that

$$x \geq \sup_{\tau \leq N} \mathbf{E}^* \mathcal{E}_\tau^{-1}(U) f_\tau,$$

where the supremum is over all stopping times $\tau \leq N$.

If in addition π is a minimal hedge, then

$$(5.2) \qquad x = \sup_{\tau \leq N} \mathbf{E}^* \, \mathcal{E}_\tau^{-1}(U) f_\tau.$$

Assume now that the initial value x and the contingent claim (f, N) satisfy (5.2). We show that in this case there exists a minimal American (x, f, N)-hedge.

To do this we proceed as follows. Let

$$Y_n = \sup_{n \leq \tau \leq N} \mathbf{E}^* \left(\frac{f_\tau}{B_\tau} \,\middle|\, \mathcal{F}_n \right),$$

where the supremum is over all stopping times $n \leq \tau \leq N$, and hence over a finite set since Ω is finite. Therefore, the variables Y_n are $\mathcal{F}_n$-measurable.

We turn to the structure of the stochastic sequence $Y = (Y_n, \mathcal{F}_n)_{n \leq N}$ and discover (see, in particular, Lecture 2, § 8) that

$$(5.3) \qquad Y_N = \frac{f_N}{B_N},$$

$$(5.4) \qquad Y_n = \max \left\{ \frac{f_n}{B_n}, \, \mathbf{E}^* \left(Y_{n+1} \,\middle|\, \mathcal{F}_n \right) \right\}$$

for $n = 0, 1, \ldots, N - 1$.

From (5.3) and (5.4) it follows directly that ($\mathbf{P}^*$-a.s.)

$$(5.5) \qquad Y_n \geq \frac{f_n}{B_n} \quad \text{for all} \quad n \leq N$$

and

$$(5.6) \qquad Y_n \geq \mathbf{E}^* \left(Y_{n+1} \,\middle|\, \mathcal{F}_n \right) \quad \text{for} \quad n \leq N - 1.$$

Thus, Y_n is a supermartingale with respect to $\mathbf{P}^*$ and majorizes the quotient f_n / B_n; moreover, it is the *smallest supermartingale* with the indicated properties (5.5) and (5.6) (*the Snell envelope*). According to the general theory of optimal stopping rules (see, in particular, Theorem 2.6), the stopping time

$$\tau_n^* = \min \left\{ n \leq k \leq N : Y_k = \frac{f_k}{B_k} \right\}$$

is optimal in the class of stopping times $n \leq \tau \leq N$:

$$\mathbf{E}^* \left(\frac{f_{\tau_n^*}}{B_{\tau_n^*}} \,\middle|\, \mathcal{F}_n \right) = \sup_{n \leq \tau \leq N} \mathbf{E}^* \left(\frac{f_\tau}{B_\tau} \,\middle|\, \mathcal{F}_n \right).$$

In particular, since $\mathcal{F}_0 = (\Omega, \varnothing)$, we have for $\tau^* = \tau_0^*$ that

$$Y_0 = \sup_{0 \leq \tau \leq N} \mathbf{E}^* \frac{f_\tau}{B_\tau} = \mathbf{E}^* \frac{f_{\tau^*}}{B_{\tau^*}}.$$

Further, according to the Doob decomposition for supermartingales (see Theorem 2.1), we have that

$$(5.7) \qquad Y_n = M_n - A_n \quad (\mathbf{P}^*\text{-a.s.}),$$

where $M = (M_n)_{n \leq N}$ $(M_0 = Y_0)$ is a martingale, and $A = (A_n)_{n \leq N}$ $(A_0 = 0)$ is a predictable nondecreasing stochastic sequence.

By the completeness of the (B, S)-market (3.1)–(3.2), the martingale M admits a representation (3.12'), which we write in the form (see the proof of Theorem 4.1)

$$(5.8) \qquad M_n = M_0 + \sum_{k=1}^{n} \frac{\gamma_k^* S_{k-1}}{B_k}(\beta_k - r_k),$$

where the γ_k^* are $\mathcal{F}_{k-1}$-measurable random variables.

Substituting (5.8) in (5.7), we get that

$$(5.9) \qquad Y_n = Y_0 + \sum_{k=1}^{n} \frac{\gamma_k^* S_{k-1}}{B_k}(\beta_k - r_k) - A_n;$$

furthermore, $Y_n \leq M_n$, $n \leq N$, and

$$Y_0 = M_0 = \sup_{0 \leq \tau \leq N} \mathbf{E}^* \frac{f_\tau}{B_\tau}.$$

Starting from M_0, γ_n^*, $n = 1, \ldots, N$, we form by analogy with the constructions of Theorem 4.1 the portfolio

$$\pi^* = (\beta_n^*, \gamma_n^*)_{n \leq N}$$

with the initial value

$$x = \sup_{0 \leq \tau \leq N} \mathbf{E}^* \mathcal{E}_\tau^{-1}(U) f_\tau.$$

Its discounted value

$$M_n^{\pi^*} = X_n^{\pi^*}/B_n$$

has the form

$$(5.10) \qquad M_n^{\pi^*} = M_0^{\pi^*} + \sum_{k=1}^{n} \frac{\gamma_k^* S_{k-1}}{B_k}(\rho_k - r_k).$$

It is immediately clear from a comparison of (5.8) and (5.10) that

$$M_n^{\pi^*} = M_n^{\pi} \quad \text{for all} \quad n \leq N.$$

Consequently,

$$(5.11) \qquad \begin{aligned} X_n^{\pi^*} &= M_n^{\pi^*} B_n = M_n B_n = (Y_n + A_n) B_n \\ &\geq Y_n B_n = \sup_{n \leq \tau \leq N} \mathbf{E}^* \left(\frac{f_\tau}{B_\tau} \,\middle|\, \mathcal{F}_n \right) B_n \\ &= \sup_{n \leq \tau \leq N} \mathbf{E}^* \left(\mathcal{E}_\tau^{-1}(U)\, \mathcal{E}_n(U)\, f_\tau \,\middle|\, \mathcal{F}_n \right). \end{aligned}$$

It follows directly from (5.11) that $X_n^{\pi^*} \geq f_n$ for $n = 0, \ldots, N$. Moreover,

$$X_0^{\pi^*} = \sup_{0 \leq \tau \leq N} \mathbf{E}^* \mathcal{E}_\tau^{-1}(U) f_\tau$$

by virtue of the equality $x/B_0 = M_0 = Y_0$. Consequently, $\pi^* \in \Pi(x, f, N)$. It is also clear that on the stochastic interval $[\![0, \tau^*[\![$

$$Y_n(\omega) = \mathbf{E}^*(Y_{n+1} \,|\, \mathcal{F}_n).$$

If we now use the fact that on this stochastic interval $A_n(\omega) = 0$, then we get from (5.9) or (5.11) that on the same interval $[\![0, \tau^*[\![$

$$X_n^{\pi^*}(\omega) = \sup_{n \leq \tau \leq N} \mathbf{E}^* \big(\mathcal{E}_\tau^{-1}(U)\, \mathcal{E}_n(U)\, f_\tau \,\big|\, \mathcal{F}_n \big).$$

From the definition of τ^*, the martingale property of Y on $[\![0, \tau^*[\![$, and the representation $A_n = \sum_{k=1}^n (Y_{k-1} - \mathbf{E}^*(Y_k \,|\, \mathcal{F}_{k-1}))$ it follows that $A_{\tau^*} = 0$. Consequently,

$$X_{\tau^*}^{\pi^*}(\omega) = Y_{\tau^*}(\omega)\, B_{\tau^*}(\omega) = f_{\tau^*}(\omega).$$

Thus, the hedge constructed is minimal.

We formulate the following fundamental result on pricing American options.

THEOREM 5.1. *Suppose that an American option with (dynamic) contingent claim* (f, N) *is traded on a complete no-arbitrage market* (3.1)–(3.2) *with deterministic sequence* $r_n > -1$, $n = 0, 1, \ldots, N$, *and martingale measure* $\mathbf{P}^*$. *Then*

1) the fair price $\mathbb{C}(N)$ *of the option is determined as the value of the optimal stopping problem*

$$(5.12) \qquad \mathbb{C}(N) = \sup_{0 \leq \tau \leq N} \mathbf{E}^* \mathcal{E}_\tau^{-1}(U) f_\tau,$$

and the stopping time $\tau^* = \tau_0^*$ *is optimal in the sense that* $\mathbf{E}^* \mathcal{E}_{\tau^*}^{-1}(U) f_{\tau^*} = \mathbb{C}(N)$;

2) the set $\Pi(\mathbb{C}(N), f, N)$ *is nonempty and consists of minimal hedges.*

The first assertion of the theorem and the nonemptiness of $\Pi(C(N), f, N)$ have already been *proved*. Further, the search for an *optimal* stopping time τ^* is inseparably connected with the problem (5.12), and the stopping time is *rational* in the sense that for any portfolio $\pi \in \mathrm{SF}$ with initial value $\mathbb{C}(N)$ and with the property

$$X_{\tau^*}^\pi(\omega) \geq f_{\tau^*}(\omega), \qquad \omega \in \Omega,$$

we have the equality

$$X_{\tau^*}^\pi(\omega) = f_{\tau^*}(\omega), \qquad \omega \in \Omega.$$

Just such a stopping time is $\tau^* = \tau_0^*$.

Indeed, if π^* is a $(\mathbb{C}(N), f, N)$-hedge and τ^* is a rational time for exercising the option, then $X_{\tau^*}^{\pi^*} = f_{\tau^*}$, and by (5.1)

$$\mathbb{C}(N) = X_0^{\pi^*} = \mathbf{E}^* \mathcal{E}_{\tau^*}^{-1}(U) X_{\tau^*}^{\pi^*} = \mathbf{E}^* \mathcal{E}_{\tau^*}^{-1}(U)\, f_{\tau^*}.$$

Conversely, if σ is a stopping time with the property

$$\mathbf{E}^* \mathcal{E}_\sigma^{-1}(U)\, f_\sigma = \sup_{\tau \leq N} \mathbf{E}^* \mathcal{E}_\tau^{-1}(U)\, f_\tau,$$

$\pi \in \mathrm{SF}$ has initial value $\mathbb{C}(N)$, and the value X_σ^π is $\geq f_\sigma$, then in view of (5.1)

$$X_0^\pi = \mathbf{E}^* \mathcal{E}_\sigma^{-1}(U)\, X_\sigma^\pi \geq \mathbf{E}^* \mathcal{E}_\sigma^{-1}(U)\, f_\sigma$$
$$= \sup_\tau \mathbf{E}^* \mathcal{E}_\tau^{-1}(U)\, f_\tau = \mathbb{C}(N).$$

Hence, $\mathbf{P}^* \{ X_\sigma^\pi > f_\sigma \} = 0$, and the stopping time σ is rational.

To *prove* the second assertion we assume that π^* is a $(C(N), f, N)$-hedge. Then by the preceding arguments, an *optimal* stopping time τ^* for the problem (5.12) is *rational*. Consequently, $X_{\tau^*}^{\pi^*} = f_{\tau^*}$, and therefore π^*-is minimal. The theorem is proved.

The introduction of the concept of a *rational* exercise time in the pricing of American options is justified by the following arguments. If, after paying the premium $\mathbb{C}(N)$, the holder exercises the option at the time τ, then he gets an amount f_τ. However, on the (B, S)-market there may exist a strategy $\pi \in$ SF such that $X_0^\pi = \mathbb{C}(N)$, and $X_\tau^\pi > f_\tau$. The holder of the option does gain, but could have received a greater return if he himself had "created" the capital X_τ^π. Consequently, his behavior on the market turns out to be *irrational*.

§3. The methodology of pricing American options

As was only just shown, the essence of pricing for American options amounts to reducing the pricing problem to a suitable optimal stopping problem. We recall that for a nonnegative stochastic sequence $X = (X_n)_{n \in \mathbb{Z}_+}$ it is necessary to find the value $V = \sup_\tau \mathbf{E} X_\tau$ and an (optimal) stopping time τ^* such that

$$(5.13) \qquad\qquad V = \mathbf{E} X_{\tau^*},$$

where the supremum is over some set of stopping times.

It is usually complicated and laborious enough to solve problems of the type (5.13) concretely, that is, to find the value V and the stopping time τ^*. Nevertheless, it is sometimes possible to use the following elegant method. The idea is to find another stochastic sequence $Y = (Y_n)_{n \in \mathbb{Z}_+}$, a function $g = g(y)$ with a unique maximum point y^*, and a positive martingale $M = (M_n)_{n \in \mathbb{Z}_+}$, $M_0 = 1$, such that for all $n \in \mathbb{Z}_+$

$$(5.14) \qquad\qquad X_n = g(Y_n) M_n.$$

It is clear from (5.14) that for all $n \in \mathbb{Z}_+$

$$X_n \le g(y^*) M_n.$$

Consequently, for any finite stopping time τ we have the relations

$$X_\tau \le g(y^*) M_\tau,$$
$$\mathbf{E} X_\tau \le g(y^*) \mathbf{E} M_\tau = g(y^*).$$

Therefore, under our assumptions $g(y^*)$ turns out to be an upper bound for the value V in the problem (5.13).

The next step is to produce a stopping time τ^* such that

$$\mathbf{E} X_{\tau^*} = \mathbf{E} g(Y_{\tau^*}) M_{\tau^*} = g(y^*).$$

Both an optimal stopping time τ^* and the value $V = g(y^*)$ itself are found in this way, and usually

$$\tau^* = \inf\{n \colon Y_n = y^*\}.$$

We realize this approach for the pricing of an American option to buy with dynamic contingent claim $f_n = (S_n - K)^+$, $n \le N$, in the framework of the following *symmetric binomial model* of a complete (B, S)-market:

$$(5.15) \qquad \begin{aligned} \Delta B_n &= r B_{n-1}, & B_0 &> 0, \\ \Delta S_n &= \rho_n S_{n-1}, & S_0 &> 0, \quad n \le N. \end{aligned}$$

With respect to the parameters of the model it is assumed that

$$a = \lambda^{-1} - 1, \quad b = \lambda - 1, \qquad \lambda > 1.$$

We remark that in this case the evolution of the prices of shares

$$S_n = S_0 \, \lambda^{\varepsilon_1 + \cdots + \varepsilon_n}$$

is a symmetric geometric random walk. According to (5.13) or (5.12), computing the fair price of this option amounts to finding

$$\mathbb{C}(N) = \sup_{\tau \leq N} \mathbf{E}^* \, \mathcal{E}_\tau^{-1}(U) \, (S_\tau - K)^+$$
$$= \sup_{\tau \leq N} \mathbf{E}^* \, (1 + r)^{-\tau} \, (S_\tau - K)^+,$$

where $\mathbf{E}^*$ is the averaging with respect to the martingale measure $\mathbf{P}^*$ corresponding to the parameter $p^* = (r - a)/(b - a)$.

For simplicity assume that $B_0 = S_0 = 1$.

Next, we have

(5.16) $$\mathcal{E}_n^{-1}(U) \, (S_n - K)^+ = (S_n - K)^+ S_n^{-1} \, \mathcal{E}_n^{-1}(U) \, S_n.$$

As we saw earlier, the sequence $(\mathcal{E}_n^{-1}(U) \, S_n)_{n \leq N}$ is a martingale with respect to the measure $\mathbf{P}^*$, and we denote it by $M = (M_n)_{n \leq N}$, $M_0 = 1$.

Keeping (5.14) in mind, we consider the nondecreasing function $g(y) = (y - K)^+ y^{-1}$. Suppose that $K = \lambda^{N_k}$, $|N_k| \leq N$. Then on the lattice λ^n, $n = 0, \pm 1, \ldots, \pm N$, the maximal value of the function g is attained at the point $y^* = \lambda^N$ and is equal to

$$g(y^*) = (\lambda^N - \lambda^{N_k})^+ \lambda^{-N}.$$

From this the choice is clear for an optimal stopping time:

$$\tau^* = \inf \{n \leq N \colon S_n = \lambda^N\},$$

that is, $\tau^* = N$.

Thus, the American option to buy coincides with a European option with exercise date N.

Problems

5.1. For a one-period binomial *symmetric* model of a (B, S)-market let us consider an American option with contingent claim $f_n = \beta^n \, (S_n - 1)^+$, $\beta \in (0, 1)$, $n = 0, 1$. Prove that:

a) for $S_0 = 1 = \lambda^0$ the fair price $\mathbb{C}(1)$ is equal to $\alpha \beta p^* \, (\lambda - 1)$, and the exercise time τ^* is identically equal to 1, where

$$\lambda > 1, \quad \alpha = \frac{1}{1 + r}, \quad p^* = \frac{r - a}{b - a} = \frac{r - (\lambda^{-1} - 1)}{\lambda - 1 - (\lambda^{-1} - 1)};$$

b) for $S_0 = \lambda$ and $\beta \in (\frac{\lambda - 1}{\lambda - \alpha}, 1)$ the fair price $\mathbb{C}(1)$ is equal to $\beta \, (\lambda - \alpha)$, and the exercise time τ^* is identically equal to 0.

5.2. Show that for the option in Problem 5.1 in the general case when $S_0 = \lambda^k$ with $k > 1$ or $k \leq -1$, the fair price $\mathbb{C}(1)$ is equal to $\max \{\lambda^k - 1, \beta \, (\lambda^k - \alpha)\}$ or 0, respectively. Further, the exercise time τ^* is equal to $\min \{0 \leq m \leq 1 : S_m \in [\lambda^{k_1^* - m}, \infty)\}$, where $k_0^* = -\infty$, $k_1^* = \max \{0, \log_\lambda \frac{1 - \alpha}{1 - \beta}\}$.

> The sixth lecture introduces G-financing strategies, whose values
> are formed with consumption or refinancing taken into account.
> Pricing of European and American options is covered, and it is
> shown that in the latter case the pricing problem reduces to an
> optimal stopping problem with a payment (see [12] and [14]).

Financial computations on a complete market with the use of nonself-financing strategies

§ 1. G-financing strategies

Continuing our study of a market (3.1)–(3.2), we can imagine a more realistic situation when a change in the portfolio is accompanied by either an inflow or an outflow of capital. Modeling this with the help of some stochastic sequence $G = (G_n)$, we introduce the class GF of (G-*financing*) strategies $\pi = (\beta_n, \gamma_n)$ such that

$$(6.1) \qquad B_{n-1}\,\Delta\beta_n + S_{n-1}\,\Delta\gamma_n = -\Delta G_n,$$

where

$$G_n = \sum_{k=1}^{n} \Delta G_k, \qquad G_0 = 0.$$

If $\Delta G \geq 0$ (respectively, $\Delta G \leq 0$), then a G-financing strategy π will be called a *strategy with consumption* (respectively, a *strategy with refinancing* or *investment*).

We remark that in view of (3.3) self-financing means 0-*financing*.

In correspondence with the "balance equation" (6.1) for the value X^π of a strategy $\pi \in$ GF we have that

$$(6.2) \qquad \begin{aligned} X_n^\pi &= \beta_n B_n + \gamma_n S_n, \\ X_{n-1}^\pi &= \beta_n B_{n-1} + \gamma_n S_{n-1} + \Delta G_n, \\ \Delta X_n^\pi &= \beta_n\,\Delta B_n + \gamma_n\,\Delta S_n - \Delta G_n. \end{aligned}$$

We get from the relations (6.2), in a way similar to that for (3.5), that for $n \leq N$

$$(6.3) \qquad \Delta X_n^\pi = r_n X_{n-1}^\pi + \gamma_n S_{n-1}(\rho_n - r_n) - (1 + r_n)\,\Delta G_n.$$

Applying the formula (2.9) to the nonhomogeneous linear stochastic equation (6.3), for the solution we get the expression

$$(6.4) \qquad X_n^\pi = \mathcal{E}_n(U) \left\{ X_0^\pi + \sum_{k=1}^n \mathcal{E}_k^{-1}(U)\, \gamma_k\, S_{k-1}\, (\rho_k - r_k) \right.$$
$$\left. - \sum_{k=1}^n \mathcal{E}_{k-1}^{-1}(U)\, \Delta G_k \right\}.$$

Setting

$$(6.5) \qquad M_n^\pi = X_0^\pi + \sum_{k=1}^n \mathcal{E}_k^{-1}(U)\, \gamma_k\, S_{k-1}\, (\rho_k - r_k),$$

$$G_n^{\mathcal{E}} = \sum_{k=1}^n \mathcal{E}_{k-1}^{-1}(U)\, \Delta G_k, \qquad G_0^{\mathcal{E}} = 0,$$

we get from (6.4) that

$$(6.6) \qquad \mathcal{E}_n^{-1}(U)\, X_n^\pi = M_n^\pi - G_n^{\mathcal{E}}.$$

From (6.5), (6.6), Theorem 3.1, and the properties of stochastic integrals with respect to martingales it follows, in particular, that the quotient $\mathcal{E}_n^{-1}(U)\, X_n^\pi$ is a *martingale* with respect to the martingale measure $\mathbf{P}^*$ if G is a martingale.

As a consequence of (6.6),

$$(6.7) \qquad \mathbf{E}^* \mathcal{E}_N^{-1}(U)\, X_N^\pi = X_0^\pi - \sum_{k=1}^N \mathbf{E}^* \mathcal{E}_{k-1}^{-1}(U)\, \Delta G_k.$$

For a (B,S)-market (3.1)–(3.2) with a specified contingent claim (f, N) in it, Definitions 4.1, 4.2, 5.1, and 5.2 still apply to a European or American option, with the class SF replaced by GF. The corresponding prices and hedges will then be called G-prices and G-hedges.

As established in Lectures 4 and 5, under conditions of a complete (B,S)-market (with unique martingale measure $\mathbf{P}^*$ and predictable sequence $r_n > -1$) the problem of pricing and hedging options has been adequately solved in the class SF.

In the next two sections an analogous problem is solved for the class of G-financing strategies (for definiteness, *strategies with consumption*). Further, *the difference between the fair price and the G-price* is clarified.

§2. Pricing European options with the use of G-financing strategies

Suppose that the G-financing strategy π is a European (x, f, N)-hedge (a G-hedge). Then it follows from (6.7) that

$$x = X_0 \geq \mathbf{E}^* \mathcal{E}_N^{-1}(U)\, f + \sum_{k=1}^N \mathbf{E}^* \mathcal{E}_{k-1}^{-1}(U)\, \Delta G_k,$$

and, if it is minimal,

$$(6.8) \qquad x = \mathbf{E}^* \left\{ \mathcal{E}_N^{-1}(U)\, f + \sum_{k=1}^N \mathbf{E}^* \mathcal{E}_{k-1}^{-1}(U)\, \Delta G_k \right\}.$$

We now show that if (6.8) holds, then there is a minimal G-hedge. To prove this assertion we define a martingale M^* by

$$(6.9) \qquad M_n^* = B_0^{-1} \mathbf{E}^* \left(\mathcal{E}_N^{-1}(U) f + \sum_{k=1}^{N} \mathcal{E}_{k-1}^{-1}(U) \Delta G_k \,\middle|\, \mathcal{F}_n \right).$$

By (6.8),

$$(6.10) \qquad M_0^* = \frac{x}{B_0}.$$

Further, since the market is complete, this martingale has the martingale representation, which we write with the help of (6.10) in the form

$$(6.11) \qquad M_n^* = \frac{x}{B_0} + \sum_{k=1}^{n} \alpha_k^* (\rho_k - r_k), \qquad n \le N.$$

We show that, knowing α_n^* and ΔG_n, we can construct a minimal (x, f, N)-hedge (a G-hedge) $\pi^* \in \mathrm{GF}$ such that the stochastic sequence

$$M_n^{\pi^*} = \left(\mathcal{E}_n^{-1}(U) X_n^{\pi^*} + \sum_{k=1}^{n} \mathcal{E}_{k-1}^{-1}(U) \Delta G_k \right) B_0^{-1}$$

is a martingale with respect to $\mathbf{P}^*$ that coincides with M^*.

In the first step we set

$$\gamma_1^* = \alpha_1^* B_1 S_0^{-1}, \qquad \beta_1^* = \frac{x - \Delta G_1 - \gamma_1^* S_0}{B_0}.$$

Then by (6.11),

$$\begin{aligned}
M_1^{\pi^*} &= B_1^{-1} X_1^{\pi^*} + B_0^{-1} \Delta G_1 \\
&= B_1^{-1} (\beta_1^* B_1 + \gamma_1^* S_1) + B_0^{-1} \Delta G_1 \\
&= \beta_1^* + \gamma_1^* B_1^{-1} S_1 + B_0^{-1} \Delta G_1 \\
&= \frac{x}{B_0} - B_0^{-1} \Delta G_1 - \alpha_1^* \frac{B_1}{B_0} + \alpha_1^* \frac{S_1}{S_0} + B_0^{-1} \Delta G_1 \\
&= \frac{x}{B_0} + \alpha_1^* (\beta_1 - r_1) = M_1.
\end{aligned}$$

Continuing this process, we get a G-financing strategy π^* such that for any $n \le N$

$$M_n^* = M_n^{\pi^*}.$$

As a result, we have with the use of (6.9) that

(6.12) $$X_n^{\pi^*} = M_n^* B_n - B_n \sum_{k=1}^n B_0^{-1} \mathcal{E}_{k-1}^{-1}(U) \, \Delta G_k$$

$$= B_n \, \mathbf{E}^* \left(B_N^{-1} f + \sum_{k=1}^N B_0^{-1} \mathcal{E}_{k-1}^{-1}(U) \, \Delta G_k \, \bigg| \, \mathcal{F}_n \right)$$

$$- B_n \sum_{k=1}^n B_0^{-1} \mathcal{E}_{k-1}^{-1}(U) \, \Delta G_k$$

$$= \mathbf{E}^* \left(\mathcal{E}_N^{-1}(U) \, \mathcal{E}_n(U) \, f + \sum_{k=n+1}^N \mathcal{E}_n(U) \, \mathcal{E}_{k-1}^{-1}(U) \, \Delta G_k \, \bigg| \, \mathcal{F}_n \right).$$

In particular, for all ω

$$X_N^{\pi^*} = f(\omega) \quad \text{and} \quad X_0^{\pi^*} = x,$$

and hence π^* is a minimal G-hedge.

Thus, we have obtained the following theorem.

THEOREM 6.1. *Consider a complete market* (3.1)–(3.2) *with deterministic sequence* $r_k > -1$ *and a martingale measure* $\mathbf{P}^*$. *Let the stochastic sequence* $G = (G_n)_{n \le N}$, $G_0 = 0$, *determine G-financing strategies on this market. Then the following assertions hold for a European option with contingent claim* (f, N):

1) *the fair price of the option is*

(6.13) $$\mathbb{C}(N, G) = \mathbf{E}^* \left(\mathcal{E}_N^{-1}(U) \, f + \sum_{k=1}^N \mathcal{E}_{k-1}^{-1}(U) \, \Delta G_k \right);$$

2) *there exists a minimal* $(\mathbb{C}, f, N)$-*hedge* $\pi^* = ((\beta_n^*, \gamma_n^*))_{n \le N}$, *and it is determined by the formulas*

$$\gamma_n^* = \frac{\alpha_n^* B_n}{S_{n-1}}, \quad \beta_n^* = \frac{X_{n-1}^{\pi^*} - \gamma_n^* S_{n-1} - \Delta G_n}{B_{n-1}},$$

where α_n^* comes from the decomposition (6.11);

3) *the value of a minimal G-hedge is determined by the formulas* (6.12).

EXAMPLE 6.1. We consider a binomial market (3.1)–(3.2) and a predictable sequence $G_n = c \sum_{k=1}^n S_{k-1}$, where c is a nonnegative constant. Let $f = f(S_N)$ be the contingent claim of the option, whose ownership ensures a dividend return equal to $c \, S_{n-1}$ at each time $n = 1, 2, \dots, N$.

It is not hard to verify that

(6.14) $$\mathbf{E}^* \left(\sum_{k=n+1}^N \mathcal{E}_n(U) \, \mathcal{E}_{k-1}^{-1}(U) \, \Delta G_k \, \bigg| \, \mathcal{F}_n \right)$$

$$= c \, \mathbf{E}^* \left(S_n + (1+r)^{-1} S_{n+1} + \cdots + (1+r)^{n+1-N} S_{N-1} \, \big| \, \mathcal{F}_n \right)$$

$$= c \, (N - n) \, S_n.$$

Consequently, according to (6.12)–(6.14), we have the following formula for the G-price:
$$\mathbb{C}(N,G) = (1+r)^{-N} F_N(S_0; p^*) + c\,N S_0,$$
where $F_N(x; p)$ is determined by the formula (4.26).

§3. Pricing American options

Suppose that a G-financing strategy π is an (x, f, N)-hedge of an American option with dynamic contingent claim $f = (f_n)_{n \leq N}$.

For an arbitrary stopping time $\tau \leq N$ we get from the formulas (6.6) that
$$\mathbf{E}^* \mathcal{E}_\tau^{-1}(U) X_\tau^\pi = X_0 - \mathbf{E}^* \sum_{k=1}^{\tau} \mathcal{E}_{k-1}^{-1}(U)\,\Delta G_k.$$

Consequently,
$$x = X_0 \geq \sup_{0 \leq \tau \leq N} \mathbf{E}^*\Big[\mathcal{E}_\tau^{-1}(U) f_\tau + G_\tau^{\mathcal{E}}\Big],$$
and for a minimal hedge
$$(6.15) \qquad x = \sup_{0 \leq \tau \leq N} \mathbf{E}^*\Big[\mathcal{E}_\tau^{-1}(U) f_\tau + G_\tau^{\mathcal{E}}\Big].$$

We assume that the initial value $x > 0$ and the dynamic contingent claim (f, N) satisfy (6.15). Let us show that in this case there is a minimal American (x, f, N)-hedge $\pi^* \in \mathrm{GF}$.

We define the stochastic sequence
$$Y_n = \sup_{n \leq \tau \leq N} \mathbf{E}^* \big(\mathcal{E}_\tau^{-1}(U) f_\tau + G_\tau^{\mathcal{E}} \,\big|\, \mathcal{F}_n\big), \qquad n \leq N.$$

Using "backward induction" in this case, we get that Y has the structure
$$(6.16) \qquad Y_N = \mathcal{E}_N^{-1}(U) f_N + G_N^{\mathcal{E}},$$
$$(6.17) \qquad Y_n = \max\big\{\mathcal{E}_n^{-1}(U) f_n + G_n^{\mathcal{E}},\ \mathbf{E}^*(Y_{n+1} \,|\, \mathcal{F}_n)\big\},$$
where $n = 0, 1, \ldots, N-1$.

It is clear from (6.16) and (6.17) that ($\mathbf{P}^*$-a.s.)
$$Y_N \geq \mathcal{E}_N^{-1}(U) f_N + G_n^{\mathcal{E}}, \qquad n \leq N,$$
$$Y_n \geq \mathbf{E}^*(Y_{n+1} \,|\, \mathcal{F}_n), \qquad n \leq N-1.$$

Consequently, we have constructed (cf. (5.5) and (5.6)) a supermartingale Y majorizing
$$\mathcal{E}_n^{-1}(U) f_n + G_n^{\mathcal{E}}.$$

According to the analogue of Theorem 2.6, the stopping time
$$\tau_n^* = \min\big\{n \leq k \leq N : Y_k = \mathcal{E}_k^{-1}(U) f_k + G_k^{\mathcal{E}}\big\}$$
is optimal among the stopping times $n \leq \tau \leq N$ in the following sense:
$$\mathbf{E}^* \big(\mathcal{E}_{\tau_n^*}^{-1}(U) f_{\tau_n^*} + G_{\tau_n^*}^{\mathcal{E}} \,\big|\, \mathcal{F}_n\big) = \sup_{n \leq \tau \leq N} \mathbf{E}^* \big(\mathcal{E}_\tau^{-1}(U) f_\tau + G_\tau \,|\, \mathcal{F}_n\big).$$

In particular, for $\tau^* = \tau_0^*$ it follows from $\mathcal{F}_0 = \{\varnothing, \Omega\}$ that

$$Y_0 = \sup_{0 \le \tau \le N} \mathbf{E}^* \left[\mathcal{E}_\tau^{-1}(U) f_\tau + G_\tau^{\mathcal{E}} \right].$$

Applying the Doob decomposition (Theorem 2.1) to the supermartingale Y, we now get that for $n = 0, \ldots, N$

$$(6.18) \qquad\qquad Y_n = M_n - A_n \quad (\mathbf{P}^*\text{-a.s.}),$$

where $M = (M_n)_{n \le N}$ $(M_0 = Y_0)$ is a martingale, and $A = (A_n)_{n \le N}$ $(A_0 = 0)$ is a predictable nondecreasing sequence.

Since the market is complete, the martingale M has the representation (6.11). Using this representation, we rewrite (6.18) in the form

$$(6.19) \qquad\qquad Y_n = x + \sum_{k=1}^{n} \alpha_k^* (\rho_k - r_k) - A_n.$$

Using x, ΔG_n, and α_n^*, we construct an (x, f, N)-hedge $\pi^* = (\beta_n^*, \gamma_n^*)_{n \le N} \in \mathrm{GF}$ with initial value x determined by the formula (6.15) such that the martingale M^{π^*} in (6.6) coincides with M: for $n = 0, \ldots, N$

$$(6.20) \qquad\qquad M_n^{\pi^*} = M_n \quad (\mathbf{P}^*\text{-a.s.}).$$

From (6.19) and (6.20) we get that

(6.21)

$$\begin{aligned}
X_n^{\pi^*} &= \mathcal{E}_n(U) \, M_n^{\pi^*} - \mathcal{E}_n(U) \, G_n^{\mathcal{E}} \\
&= \mathcal{E}_n(U) \, (Y_n + A_n) - \mathcal{E}_n(U) \, G_n^{\mathcal{E}} \\
&\ge \mathcal{E}_n(U) \, Y_n - \mathcal{E}_n(U) \, G_n^{\mathcal{E}} \\
&= \mathcal{E}_n(U) \sup_{n \le \tau \le N} \mathbf{E}^* \left(\mathcal{E}_\tau^{-1}(U) f_\tau + G_\tau^{\mathcal{E}} \,\big|\, \mathcal{F}_n \right) - \mathcal{E}_n(U) \, G_n^{\mathcal{E}} \\
&= \sup_{n \le \tau \le N} \mathbf{E}^* \left(\mathcal{E}_n(U) \, \mathcal{E}_\tau^{-1}(U) f_\tau + \mathcal{E}_n(U) \, G_\tau^{\mathcal{E}} \,\big|\, \mathcal{F}_n \right) - \mathcal{E}_n(U) \, G_n^{\mathcal{E}}.
\end{aligned}$$

From (6.21) with $\tau \equiv n$ it follows that for all $n \le N$

$$X_n^{\pi^*} \ge f_n \quad (\mathbf{P}^*\text{-a.s.}),$$

and hence π is an (x, f, N)-hedge.

Its minimality, and also the rationality of the stopping time $\tau^* = \tau_0^*$, are proved by the same arguments as in the proof of Theorem 5.1.

The foregoing can be summarized as follows.

THEOREM 6.2. *The fair price of an American option with contingent claim (f, N) on a complete (B, S)-market (3.1)–(3.2) with deterministic sequence $r_n > -1$ and with G-financing recomputation of the value (6.2) is a solution of the following optimal stopping problem with a payment:*

$$\mathbb{C}(N) = \sup_{0 \le \tau \le N} \mathbf{E}^* \left(\mathcal{E}_\tau^{-1}(U) f_\tau + G_\tau^{\mathcal{E}} \right),$$

where $\mathbf{E}^*$ is the averaging with respect to the (unique) martingale measure $\mathbf{P}^*$.

PROBLEMS

6.1. Prove the uniqueness of a minimal G-hedge (see § 6.2).

6.2. Suppose that on a complete (B, S)-market with $B_n \equiv 1$ a European option with contingent claim (f, N) is given. Ownership of the option brings in dividends $\Delta D_n \geq 0$, $n \leq N$. We define the class of G-financing strategies $\pi(G)$ such that $X_N^{\pi(G)} \geq f$ and $\mathbf{E}^* \Delta G_n \geq \mathbf{E}^* \Delta D_n$. Prove that a D-financing strategy has minimal value in this class.

LECTURE 7

The seventh lecture is devoted to the study of an incomplete market (3.1). An example is given of such a market, and a theorem is
proved on computing the ask and bid prices of a European option
(see [21], [22], and [33]). Approaches are presented to accounting
for the risk associated with the contingent claim on an option.
It is proposed to use a **P**-discounting portfolio. Further, it turns
out that this approach is equivalent to the traditional one based
on the martingale projection technique and reducing to a minimal
martingale measure (see [25]–[27], [38], [39], [16]).

Incomplete markets. Pricing of options and problems of minimizing risk

§ 1. Ask and bid prices. An example of an incomplete market

The model considered earlier of a no-arbitrage (B, S)-market (3.1)–(3.2) was en-
dowed with the property of *completeness*, which, according to Theorem 3.3, means
the *uniqueness* of a martingale measure. All financial computations have been carried out with respect to the latter. The question arises as to the possibility of
similar computations under conditions of an *incomplete market* (3.1)–(3.2), when
the martingale measure is *not unique*.

First, we make clear what is natural for us to regard as the price of an option
(of European type) with contingent claim (f, N) in the framework of an *incomplete
model* of a (B, S)-market (3.1)–(3.2). A participant in the market can appear both
as a seller and as a buyer of the option. The different goals of the seller and buyer
lead to the setting of generally *different* prices $\mathbb{C}^*(N)$ and $\mathbb{C}_*(N)$ of the seller and
buyer, and to the appearance of a nonzero difference $\mathbb{C}^* - \mathbb{C}_*$, called the *spread*.

We remark that in the case of a complete market there is the possibility of
combining the opposing interests of the seller and the buyer, as expressed in the
existence of a *fair* price $\mathbb{C}(N)$.

Let us make the above more precise. To do this we introduce the following
classes of self-financing portfolios $\pi = (\beta, \gamma)$:

$$\Pi^*(x, f, N) = \{\pi \in \mathrm{SF} \colon X_N^\pi \geq f, \ X_0^\pi = x > 0\},$$
$$\Pi_*(x, f, N) = \{\pi \in \mathrm{SF} \colon X_N^\pi \leq f, \ X_0^\pi = x > 0\}.$$

63

We now define the *ask* and *bid* prices of a contingent claim (f, N) (of an option with contingent claim (f, N)) by the formulas

$$\mathbb{C}^*(N) = \inf \{x \colon \Pi^*(x, f, N) \neq \varnothing\},$$
$$\mathbb{C}_*(N) = \sup \{x \colon \Pi_*(x, f, N) \neq \varnothing\}.$$

Let $\pi \in \mathrm{SF}$. Then by (3.5) and (3.5'), the value X^π of this portfolio is

$$X_N^\pi = \mathcal{E}_N(U) \left\{ X_0 + \sum_{k=1}^{N} \mathcal{E}_{k-1}^{-1}(U) \, \gamma_k S_{k-1} \left(\rho_k - r_k\right) \right\}.$$

Consequently, for the martingale measure $\widetilde{\mathbf{P}}$

$$\widetilde{\mathbf{E}} \, \mathcal{E}_N^{-1}(U) \, X_N^\pi = \widetilde{\mathbf{E}} \, X_0 = x.$$

Therefore,

(7.1) $$\mathbb{C}_* \leq \widetilde{\mathbf{E}} \, \mathcal{E}_N^{-1}(U) \, f \leq \mathbb{C}^*.$$

If the measure $\widetilde{\mathbf{P}}$ is unique, and the deterministic sequence r_n is such that $r_n > -1$, then by Theorem 4.1 there is a minimal $(\mathbb{C}(N), f, N)$-hedge, which clearly belongs to both the classes Π^* and Π_*, and

$$\mathbb{C}(N) = \widetilde{\mathbf{E}} \, \mathcal{E}_N^{-1}(U) \, f.$$

Thus, for a complete market we arrive at the equality

$$\mathbb{C}_* = \mathbb{C} = \mathbb{C}^*,$$

that is, *a complete market is characterized by the absence of a spread.*

But if the market is incomplete, then it follows from (7.1) that

(7.2) $$\mathbb{C}_*(N) \leq \inf_{\mathbb{P}^*} \widetilde{\mathbf{E}} \, \mathcal{E}_N^{-1}(U) \, f \leq \sup_{\mathbb{P}^*} \widetilde{\mathbf{E}} \, \mathcal{E}_N^{-1}(U) \, f \leq \mathbb{C}^*(N).$$

The inequalities (7.2) suggest the idea that in a whole series of cases the left- and right-hand inequalities in (7.2) become equalities. The supremum is usually strictly greater than the infimum, and hence *an incomplete market is characterized by a nonzero spread.*

This idea is realized below, first in the form of an example, and then in the form of a general theorem.

EXAMPLE 7.1. Consider the following one-period model of a (B, S)-market (3.1)–(3.2):

$$B_1 = B_0 = S_0 \equiv 1, \quad S_1 = 1 + \rho,$$

where ρ can take the three values $+1/2$, 0, and $-1/2$ with positive probabilities p_1, p_2, and p_3, where $p_1 + p_2 + p_3 = 1$.

The martingale property of the measure $\mathbf{P}$ induced by the triple (p_1, p_2, p_3) means that

$$\mathbf{E}\,\rho = \frac{1}{2}p_1 - \frac{1}{2}p_3 = 0,$$

and therefore $p_3 = p_1$. Further,

$$p_2 = 1 - p_1 - p_3 = 1 - 2p_1,$$

as a result, for any p_1 with $0 < p_1 < 1/2$ the formulas $p_3 = p_1$ and $p_2 = 1 - 2p_1$ determine a martingale measure. For the model under consideration we even have a whole family $\mathbb{P}^*$ of such measures, and hence the market is incomplete.

Suppose that the contingent claim $(f, 1)$ has the form

$$f = f(S_1) = (S_1 - 1)^+ = \rho^+.$$

Let us find the ask price $\mathbb{C}^*$ and the bid price $\mathbb{C}_*$ of this claim.

We consider the following two sequences of martingale measures $\mathbf{P}_n^*$ and $\mathbf{P}_*^n$:

$$\mathbf{P}_n^* \left\{ \rho = +\frac{1}{2} \right\} = \mathbf{P}_n^* \left\{ \rho = -\frac{1}{2} \right\} = \frac{1}{2} \left\{ 1 - \frac{1}{n} \right\}, \quad \mathbf{P}_n^* \left\{ \rho = 0 \right\} = \frac{1}{n};$$

$$\mathbf{P}_*^n \left\{ \rho = +\frac{1}{2} \right\} = \frac{1}{2n} = \mathbf{P}_*^n \left\{ \rho = -\frac{1}{2} \right\}, \quad \mathbf{P}_*^n \left\{ \rho = 0 \right\} = 1 - \frac{1}{n}.$$

It is clear that as $n \to \infty$

$$\mathbf{E}_n^* \rho^+ = \frac{1}{4} \left(1 - \frac{1}{n} \right) \uparrow \frac{1}{4}$$

and

$$\mathbf{E}_*^n \rho^+ = \frac{1}{4n} \downarrow 0,$$

and hence the inequalities (7.2) give us that

$$\mathbb{C}_*(1) = 0, \quad \mathbb{C}^*(1) = \frac{1}{4}.$$

We remark that the limit measures $\mathbf{P}^*$ and $\mathbf{P}_*$, which are concentrated at the respective points $\{-1/2\}$, $\{1/2\}$ and $\{0\}$, are *extremal* for $\mathbb{P}^*$, and averaging the contingent claim with respect to them yields the values of the ask and bid prices.

§2. Formulas for computing the ask and bid prices for convex contingent claims

Suppose that in the model of a (B, S)-market (3.1) the quantities ρ_n take the values $a = a_1 < a_2 < \cdots < a_m = b$, and $r_n \equiv r \in \{a_2, \ldots, a_{m-1}\}$. For the family $\mathbb{P}^*$ of martingale measures we have the relation $\mathbf{E} \rho_n = r$. We define two "almost martingale" measures $\mathbf{P}^*$ and $\mathbf{P}_*$ by

$$\mathbf{P}^* \{\rho_n = b\} = \frac{r - a}{b - a} = p^*,$$

$$\mathbf{P}^* \{\rho_n = a\} = \frac{b - r}{b - a} = 1 - p^*,$$

$$\mathbf{P}_* \{\rho_n = r\} = 1.$$

It is clear that $\mathbf{P}^* \not\sim \mathbf{P}$ and $\mathbf{P}_* \not\sim \mathbf{P}$, but $\mathbf{E}^* \rho_n = \mathbf{E}_* \rho_n = r$.

Proceeding by analogy with Example 7.1, we define sequences of *martingale measures* $\mathbf{P}_n^*$ and $\mathbf{P}_*^n$ that converge to $\mathbf{P}^*$ and $\mathbf{P}_*$, respectively (in the sense that $\mathbf{P}_n^* \{A\} \to \mathbf{P}^* \{A\}$ and $\mathbf{P}_*^n \{A\} \to \mathbf{P}_* \{A\}$ for $A = \{\rho = a_k\}$, $k = 1, \ldots, m$):

$$\mathbf{P}_n^* \{\rho = b\} = p^* - \frac{1}{2n},$$

$$\mathbf{P}_n^* \{\rho = a\} = 1 - p^* - \frac{1}{2n},$$

$$\mathbf{P}_*^n \{\rho = r\} = 1 - \frac{1}{n},$$

with the remaining probability mass $1/n$ distributed at the remaining points in such a way that $\mathbf{E}_n^* \rho = r$ and $\mathbf{E}_*^n \rho = r$.

We formulate the basic result about computing the ask and bid prices. In its proof we carry out the necessary averagings at once with respect to the measures $\mathbf{P}^*$ and $\mathbf{P}_*$ to which the sequences of martingale measures $\mathbf{P}_n^*$ and $\mathbf{P}_*^n$ converge.

THEOREM 7.1. 1) *In a finite no-arbitrage market* (3.1)–(3.2) *with some contingent claim* (f, N) *the ask and bid prices are connected by the relation* (7.2).

2) *If the contingent claim* $f = f(S_N)$ *is a convex function of* S_N, *then*

(7.3)
$$\mathbb{C}^*(N) = \sup_{\mathbb{P}^*} \widetilde{\mathbf{E}}\, \mathcal{E}_N^{-1}(U)\, f,$$
$$\mathbb{C}_*(N) = \inf_{\mathbb{P}^*} \widetilde{\mathbf{E}}\, \mathcal{E}_N^{-1}(U)\, f.$$

Further, the supremum and infimum are attained on the respective "extreme" measures $\mathbf{P}^*$ *and* $\mathbf{P}_*$ *for* $\mathbb{P}^*$.

PROOF. The first assertion was proved in a more general case in the preceding section. For the second assertion we proceed as follows. We fix the values $\rho_1, \ldots, \rho_{N-1}$ and define the quantities $\mu = \mu(\rho_1, \ldots, \rho_{N-1})$ and $\nu = \nu(\rho_1, \ldots, \rho_{N-1})$ to be the solutions of the equations

(7.4)
$$f\big(S_{N-1}(1 + b)\big) = \mu\big(S_{N-1}(1 + b)\big) + \nu,$$
$$f\big(S_{N-1}(1 + a)\big) = \mu\big(S_{N-1}(1 + a)\big) + \nu,$$

that is,

(7.5)
$$\mu = \frac{f(S_{N-1}(1 + b)) - f(S_{N-1}(1 + a))}{S_{N-1}(b - a)}$$
$$\nu = \frac{(1 + b)f(S_{N-1}(1 + a)) - (1 + a)f(S_{N-1}(1 + b))}{(b - a)}.$$

Consider the linear function

$$y(\rho_1, \ldots, \rho_{N-1}, x) = \mu(\rho_1, \ldots, \rho_{N-1})\, S_{N-1}\,(1 + x) + \nu(\rho_1, \ldots, \rho_{N-1}),$$

which, by (7.5), has value at $x = r$ equal to

(7.6)
$$\begin{aligned}
y(r) &= \mu\, S_{N-1}\,(1 + r) + \nu \\
&= \frac{1 + r}{b - a}\Big[f\big(S_{N-1}\,(1 + b)\big) - f\big(S_{N-1}\,(1 + a)\big)\Big] \\
&\quad + \frac{(1 + b)f(S_{N-1}\,(1 + a)) - (1 + a)f(S_{N-1}\,(1 + b))}{(b - a)}.
\end{aligned}$$

We remark that

(7.7)
$$\mathbf{E}^*\Big(f\big(S_{N-1}\,(1 + \rho_N)\big)\,\Big|\,\mathcal{F}_{N-1}\Big) = \frac{1}{b - a}\Big\{(r - a)f\big(S_{N-1}\,(1 + b)\big)$$
$$+ (b - r)f\big(S_{N-1}\,(1 + a)\big)\Big\}.$$

From (7.6) and (7.7),

$$\mathbf{E}^*\big(f(S_N)\,\big|\,\mathcal{F}_{N-1}\big) = \mu\, S_{N-1}\,(1 + r) + \nu.$$

Therefore, setting

$$\beta_N = \frac{\nu}{B_0(1+r)^N}\,, \qquad \gamma_N = \mu,$$

we get that

(7.8) $\beta_N B_0 (1+r)^{N-1} + \gamma_N S_{N-1} = \mathbf{E}^* \big(f(S_N)(1+r)^{-1} \,\big|\, \mathcal{F}_{N-1}\big).$

Note that, by the convexity of f,

(7.9) $y(x) \geq f\big(S_{N-1}(1+x)\big), \qquad a \leq x \leq b.$

By backward induction in the relation (7.8) we construct a self-financing strategy $\pi = (\beta_k, \gamma_k)_{k \leq N}$ such that

(7.10) $$\beta_0 B_0 + \gamma_0 S_0 = \mathbf{E}^* \frac{f(S_N)}{(1+r)^N}.$$

In view of (7.9) we have that $\pi \in \Pi^*$, and hence we get from (7.10) that

$$\mathbb{C}^*(N) = \inf_{\Pi^*} (\beta_0 B_0 + \gamma_0 S_0) \geq \sup_{\mathbb{P}^*} \widetilde{\mathbf{E}}\, \frac{f(S_N)}{(1+r)^N}.$$

This and (7.2) imply the first equality in (7.3).

To prove the second equality in (7.3) we find by the convexity of f that for fixed $\rho_1, \ldots, \rho_{N-1}$ and for all x

(7.11) $f\big(S_{N-1}(1+x)\big) \geq f\big(S_{N-1}(1+r)\big) + (x - r)\,\lambda(\rho_1, \ldots, \rho_{N-1}),$

where $\lambda = \lambda(\rho_1, \ldots, \rho_{N-1})$ is some number.

By taking the conditional expectation $\widetilde{\mathbf{E}}\,(* \,|\, \mathcal{F}_{N-1})$ with respect to $\widetilde{\mathbf{P}} \in \mathbb{P}^*$, we get from (7.11) that

(7.12) $$\widetilde{\mathbf{E}}\,\big(f(S_N)\,(1+r)^{-N} \,\big|\, \mathcal{F}_{N-1}\big)$$
$$\geq \widetilde{\mathbf{E}}\,\big(f(S_{N-1}(1+r))(1+r)^{-N} \,\big|\, \mathcal{F}_{N-1}\big)$$
$$= \mathbf{E}_* \,\big(f(S_N)(1+r)^{-N} \,\big|\, \mathcal{F}_{N-1}\big).$$

Consequently,

(7.13) $$\inf_{\mathbb{P}^*} \widetilde{\mathbf{E}}\, \frac{f(S_N)}{(1+r)^N} = \mathbf{E}_* \, \frac{f(S_N)}{(1+r)^N}.$$

Further, it follows from (7.11) that

(7.14) $f\big(S_{N-1}(1+\rho_N)\big) \geq \beta_N B_0 (1+r)^N + \gamma_N S_{N-1}(1+\rho_N),$

where

(7.15) $$\beta_N = \frac{f(S_{N-1}(1+r)) - (1+r)\lambda}{(1+r)^N},$$

(7.16) $$\gamma_N = \frac{\lambda}{S_{N-1}}.$$

From (7.12)–(7.16),

(7.17) $\beta_N B_0 (1+r)^{N-1} + \gamma_N S_{N-1} = \mathbf{E}_* \big(f(S_N)\,(1+r)^{-1} \,\big|\, \mathcal{F}_{N-1}\big).$

By backward induction in this relation we thus construct a $\pi = (\beta_k, \gamma_k)_{k \leq N} \in \Pi_*$ such that

$$(7.18) \qquad \beta_0 \, B_0 + \gamma_0 \, S_0 = \mathbf{E}_* \, \frac{f(S_N)}{(1+r)^N} \, .$$

From (7.18) and (7.13) we obtain the relation

$$\mathbb{C}_* = \sup_{\Pi_*} \left(\beta_0 \, B_0 + \gamma_0 \, S_0 \right) \geq \inf_{\mathbb{P}^*} \, \widetilde{\mathbf{E}} \, \frac{f(S_N)}{(1+r)^N} \, ,$$

and together with it the second assertion in (7.3).

The theorem is proved.

As we saw earlier, in a complete market the use of nonself-financing strategies leads to a change in the price of a contingent claim by the sum of the discounted investments and the consumption, averaged with respect to the unique martingale measure. What happens in an incomplete market?

§ 3. On financial computations taking into account the risks of hedging contingent claims

The financial computations that have been carried out so far have exploited the idea of *discounting*, that is, equalizing the cost of capital at different times. The last circumstance manifested itself in the fact that the ratio of the value of a self-financing portfolio to the amount of the bank account turned out to be a martingale with respect to a certain (martingale) measure. However, the use of a bank account is not always the only reliable and technically convenient alternative to various investments. Such a role can be played by any self-financing portfolio with positive value. In this connection the following *definition* is natural.[1]

A self-financing strategy $\varphi = (\xi, \eta)$ having a positive value $A = (A_n)_{n \in \mathbb{Z}_+}$ is called a **P**-*discounting portfolio* if for any other self-financing portfolio $\pi = (\beta, \gamma)$ with value X^π the quotient X^π / A is a martingale with respect to the original measure **P**.

We remark that a **P**-discounting portfolio is *unique*.

Indeed, if $\varphi = (\xi, \eta)$ and $\varphi' = (\xi', \eta')$ are two such portfolios with values A and A', $A_0 = A'_0 = 1$, then A/A' and A'/A are martingales with respect to **P**. Consequently, $\mathbf{E}\,(A_n/A'_n) = \mathbf{E}\,[(A_n/A'_n)^{-1}] = 1$, and in view of the convexity of the function $1/x$ and Jensen's inequality we have that $A_n/A'_n \equiv 1$.

Assuming the *predictability* of $r_n > -1$, we discuss the properties of such portfolios.

For a **P**-discounting portfolio $\varphi = (\xi, \eta)$ with value $A = (A_n)_{n \in \mathbb{Z}_+}$, $A_0 = 1$, we set $l_n = \eta_n \, S_{n-1}/A_{n-1}$. Then in view of (3.1) and the self-financing property of φ we have that

$$\begin{aligned}
\Delta A_n &= \xi_n \, \Delta B_n + \eta_n \, \Delta S_n = \xi_n \, (r_n \, B_{n-1}) + \eta_n \, (\rho_n \, S_{n-1}) \\
&= \xi_n r_n \, B_{n-1} + \eta_n r_n \, S_{n-1} + \eta_n \, S_{n-1} \rho_n - \eta_n \, S_{n-1} r_n \\
&= r_n \, A_{n-1} + l_n \, A_{n-1} \, (\rho_n - r_n) \\
&= A_{n-1} \big(r_n + l_n \, (\rho_n - r_n) \big).
\end{aligned}$$

---

[1]The definition of a **P**-discounting portfolio and the clarification of its interrelation with the concept of a minimal martingale measure (see [27], [39]) is apparently due to D. O. Kramkov.

Consequently,

(7.19)
$$A_n = \mathcal{E}_n\left(\sum_{k=1}^{n} \left(r_k + l_k\left(\rho_k - r_k \right) \right) \right),$$

and under the condition

(7.20)
$$r_n + l_n\left(\rho_n - r_n \right) > -1$$

the value A_n is positive.

Along with (7.19) and (7.20), the **P**-discounting portfolio with value A is characterized by another "martingale" property.

If $\pi = (\beta, \gamma) \in$ SF *and* X^π *is the value of this portfolio, then* X_n^π / A_n *is a martingale with respect to* **P** *if and only if*

$$M_n = \sum_{k=1}^{n} \frac{\rho_k - r_k}{1 + r_k + l_k\left(\rho_k - r_k \right)} \qquad (M_0 = 0)$$

is a martingale with respect to the same measure.

For the proof we use the formulas (3.5′) and (7.19) and we get that

(7.21)
$$\Delta\left(\frac{X^\pi}{A} \right)_n = \frac{X_n^\pi}{A_n} - \frac{X_{n-1}^\pi}{A_{n-1}}$$

$$= \frac{\mathcal{E}_n(U)}{A_n}\left[X_0 + \sum_{k=1}^{n} \gamma_k\, S_{k-1}\mathcal{E}_k^{-1}(U)\left(\rho_k - r_k \right) \right]$$

$$\quad - \frac{\mathcal{E}_{n-1}}{A_{n-1}}\left[X_0 + \sum_{k=1}^{n-1} \gamma_k\, S_{k-1}\mathcal{E}_k^{-1}(U)\left(\rho_k - r_k \right) \right]$$

$$= \frac{\mathcal{E}_{n-1}(U)}{A_{n-1}}\left[\frac{1 + r_n}{1 + r_n + l_n\left(\rho_n - r_n \right)}\left(X_0 + \sum_{k=1}^{n-1} \gamma_k\, S_{k-1}\mathcal{E}_k^{-1}(U)\left(\rho_k - r_k \right) \right) \right.$$

$$\quad + \frac{1 + r_n}{1 + r_n + l_n\left(\rho_n - r_n \right)}\gamma_n\, S_{n-1}\mathcal{E}_n^{-1}(U)\left(\rho_n - r_n \right)$$

$$\quad \left. - \left(X_0 + \sum_{k=1}^{n-1} \gamma_k\, S_{k-1}\mathcal{E}_k^{-1}(U)\left(\rho_k - r_k \right) \right) \right]$$

$$= \frac{\mathcal{E}_{n-1}(U)}{A_{n-1}}\left[-l_n \frac{\rho_n - r_n}{1 + r_n + l_n\left(\rho_n - r_n \right)} \right.$$

$$\quad \times \left(X_0 + \sum_{k=1}^{n-1} \gamma_k\, S_{k-1}\mathcal{E}_k^{-1}(U)\left(\rho_k - r_k \right) \right)$$

$$\quad \left. + \frac{1 + r_n}{1 + r_n + l_n\left(\rho_n - r_n \right)}\gamma_n\, S_{n-1}\mathcal{E}_n^{-1}(U)\left(\rho_n - r_n \right) \right]$$

$$= \frac{X_{n-1}}{A_{n-1}}\left(-l_n \frac{\rho_n - r_n}{1 + r_n + l_n\left(\rho_n - r_n \right)} \right)$$

$$\quad + \frac{\gamma_n S_{n-1}}{A_{n-1}} \frac{\rho_n - r_n}{1 + r_n + l_n\left(\rho_n - r_n \right)}.$$

It is clear that (7.21) directly implies the above assertion, and the existence of the sequence $(l_n)_{n \geq 0}$ and the martingale M determines *conditions for the existence of a* **P**-*discounting portfolio*.

We note that the bank account B is a **P**-discounting portfolio if and only if **P** is a martingale measure.

Let us now consider a *nonself-financing* portfolio $\pi = (\beta, \gamma)$ such that the change in the value X^π is determined by a sequence of investments and consumption $G_n(\pi)$: $\Delta X_n^\pi = \beta_n \Delta B_n + \gamma_n \Delta S_n - \Delta G_n(\pi)$. Similar to the case of (7.21) and (6.4), we have that

$$(7.22) \quad \Delta\left(\frac{X^\pi}{A}\right)_n = \left(\frac{\gamma_n S_{n-1}}{A_{n-1}} - l_n \frac{X_{n-1}^\pi}{A_{n-1}}\right) \frac{\rho_n - r_n}{1 + r_n + l_n (\rho_n - r_n)} + \frac{\Delta G_n}{A_n}(1 + r_n).$$

A strategy π is said to be *mean self-financing* ($\pi \in \mathrm{SF}_m$) if the stochastic sequence $G_n^A = \sum_{k=1}^n A_k^{-1} \Delta G_k (1 + r_k)$, $G_0^A = 0$, is a *martingale with respect to* **P**.

It follows from the definition and (7.22) that X is *the value of a mean self-financing portfolio if and only if* X/A *is a martingale with respect to* **P**.

If we consider a European option with contingent claim (f, N), then we define its *fair price* $\mathbb{C}_m$ to be the *minimal initial value* X_0 in the class of strategies $\pi \in \mathrm{SF}_m$ such that $X_N^\pi \geq f$. It is clear from the preceding that $\mathbb{C}_m = A_0 \, \mathbf{E} \, A_N^{-1} f$.

The use of *nonself-financing* strategies (for example, in the class SF_m) bears a certain *risk* of nonpayment of the claim (f, N). As a measure of such risk for a strategy $\pi \in \mathrm{SF}_m$ we consider the variance $R^\pi = \mathbf{E} \, (G_N^A)^2$. Then the *risk of hedging* the contingent claim (f, N) is defined to be the quantity $R = \inf_{\pi \in \mathrm{SF}_m} R^\pi$, where the strategy π has initial value $\mathbb{C}_m$.

Turning now to the formula (7.22), we see easily that the risk of hedging the contingent claim (f, N) is attained on a strategy for which the martingales M and G^A *are orthogonal*.

In *the traditional approach*, which is treated below, discounting is realized with the help of the value $B_n = \mathcal{E}_n(U)$ of the portfolio $\varphi = ((\xi_n, 0))_{n \in \mathbb{Z}_+}$ with the subsequent choice of a martingale measure $\mathbf{P}^*$, with respect to which X^π/B is a martingale for any strategy $\pi \in \mathrm{SF}$. If such a measure $\mathbf{P}^*$ is *unique* (a complete market), then all the computations are carried out by means of $\mathbf{P}^*$ and do not depend on the original measure $\mathbf{P}$.

For an *incomplete market* there can be many such measures in general, and hence many strategies that minimize risk. However, often (but not always, as examples show) there exists a single martingale measure $\widehat{\mathbf{P}}$ that is universal for all claims, and with respect to which financial computations and minimization of risk are implemented according to the same scheme as for a complete market. This measure, called a *minimal martingale measure*, is distinguished in the class $\mathbb{P}^*$ by the property that for any mean self-financing (with respect to $\widehat{\mathbf{P}}$) strategy π the martingale (with respect to $\widehat{\mathbf{P}}$) $G_n^{\mathcal{E}}(\pi) = \sum_{k=1}^n \mathcal{E}_{k-1}^{-1}(U) \Delta G_k(\pi)$, which is orthogonal to the martingale (with respect to $\widehat{\mathbf{P}}$) $M_n = \sum_{k=1}^n (\rho_k - r_k)/(1 + r_k)$, also remains a martingale with respect to the original measure $\mathbf{P}$. Furthermore, $\frac{d\widehat{\mathbf{P}}_n}{d\mathbf{P}_n} = \mathrm{const} \frac{B_n}{A_n}$.

It is important to note that *the existence of a minimal martingale measure is equivalent to the existence of a* **P**-*discounting portfolio*.

In conclusion we briefly explain how to implement pricing and hedging of a contingent claim with minimization of risk in the framework of the traditional approach. An adequate solution of this problem is given with the help of the martingale "projection" technique of Kunita and Watanabe.

Assuming that the *original measure* $\mathbf{P} = \mathbf{P}^*$ *already is a martingale measure,* we get in a way analogous to that for (6.4) that for any strategy $\pi \in \mathrm{SF}_m$

$$(7.23) \qquad \mathcal{E}_n^{-1}(U)\, X_n^\pi = X_0 + \sum_{k=1}^{n} \gamma_k\, S_{k-1} \mathcal{E}_k^{-1}(U)\,(\rho_k - r_k) + G_n^{\mathcal{E}}(\pi).$$

We define the *remaining risk* of such a strategy by the formula $R_n^\pi = \mathbf{E}\left((G_N^{\mathcal{E}} - G_n^{\mathcal{E}})^2 \mid \mathcal{F}_n\right)$.

Minimization of R_n^π is understood in the following sense. A strategy $\widetilde{\pi}$ is said to be (n, π)-*coincident* if $X_N^\pi = X_N^{\widetilde{\pi}}$, $\widetilde{\beta}_k = \beta_k$ for $k \leq n-1$, and $\widetilde{\gamma}_k = \gamma_k$ for $k \leq n$. Then the strategy π *minimizes the risk* if $R_n^\pi \leq R_n^{\widetilde{\pi}}$ (**P**-a.s.) for any n and any (n, π)-coincident strategy $\widetilde{\pi}$.

A strategy such that $G_n^{\mathcal{E}}(\pi)$ is a martingale orthogonal to M (with respect to the measure $\mathbf{P}$) is said to be *optimal.*

It turns out that these concepts are the same. For our purposes it is more convenient to use the concept of an optimal strategy.

Consider the martingale $M_n^f = \mathbf{E}\left(\mathcal{E}_N^{-1}(U)\, f \mid \mathcal{F}_n\right)$, which has Kunita-Watanabe decomposition $M_n^f = X_0 + \sum_{k=1}^{n} \gamma_k^f\, \Delta(S/\mathcal{E})_k + L_n^f$ with $X_0 = \mathbf{E}\, \mathcal{E}_N^{-1}(U)\, f$, where $S/\mathcal{E}$ and L^f are orthogonal (with respect to $\mathbf{P}$) martingales, and γ^f is a predictable sequence.

We note that $\Delta(S/\mathcal{E})_n = S_{n-1} \mathcal{E}_n^{-1}(U)\,(\rho_n - r_n)$, and hence

$$(7.24) \qquad M_n^f = X_0 + \sum_{k=1}^{n} \gamma_k^f\, S_{k-1} \mathcal{E}_k^{-1}(U)\,(\rho_k - r_k) + L_n^f.$$

Now if we set $X_n = \mathcal{E}_n(U)\, M_n^f$, $\gamma_n = \gamma_n^f$, and $\beta_n = \mathcal{E}_n^{-1}(U)\,(X_n - \gamma_n S_n)$, we get in view of (7.23) and (7.24) an *optimal* strategy $\pi = (\beta, \gamma)$ with terminal value precisely equal to the contingent claim.

EXAMPLE 7.2. We consider the one-period model (3.1) with $B_1 = B_0 = 1$ and $S_1 = S_0\,(1 + \rho)$. For the contingent claim $(f, 1)$ we construct a strategy minimizing risk. Here it suffices to know X_0 and γ_1, since β_1 is determined from the *attainability* of this contingent claim: $\beta_1 = f - \gamma_1 S_1$.

For the strategy (β_1, γ_1) the change of value has the form $\Delta X_1 = \gamma_1 \Delta S_1 + (\beta_1 - \beta_0) = \gamma_1 \Delta S_1 + \Delta G_1$.

Consequently, the *risk* from choosing such a strategy for hedging the contingent claim f is equal to $\mathbf{E}\,(\Delta G_1)^2 = \mathbf{E}\,(f - X_0 - \gamma_1 \Delta S_1)^2$.

From this it is clear that minimization of risk reduces to the well-known problem of finding a *best linear estimator* for f, which is determined by the formulas:

$$\widehat{\gamma}_1 = \frac{\mathrm{cov}\,(f, S_1)}{\mathbf{D}\, S_1} = \frac{\mathrm{cov}\,(f, \Delta S_1)}{\mathbf{D}\, \Delta S_1} = \frac{1}{S_0\, \mathbf{D}\, \rho}[\mathbf{E}\, f\rho - \mathbf{E}\, f\, \mathbf{E}\, \rho],$$

$$X_0 = \mathbf{E}\, f - \widehat{\gamma}_1\, \mathbf{E}\, \Delta S_1 = \frac{1}{\mathbf{D}\, \rho}[\mathbf{E}\, \rho^2\, \mathbf{E}\, f - \mathbf{E}\, f\rho\, \mathbf{E}\, \rho].$$

PROBLEMS

7.1. Prove that the optimal strategy coincides with the strategy minimizing risk.

7.2. On an incomplete (B, S)-market consider a portfolio $\pi \in$ SF, and denote its initial and terminal values by x and $X_N^\pi(x)$, respectively. Assuming the *martingale property* for the original measure $\mathbf{P}$, find $x^* > 0$ and $\pi^* \in$ SF such that

$$\mathbf{E}\left(X_N^{\pi^*}(x^*) - f\right)^2 = \inf_{x,\pi} \mathbf{E}\left(X_N^\pi(x) - f\right)^2,$$

where f is an $\mathcal{F}_N$-measurable contingent claim.

7.3. Prove that $[0, \mathbb{C}_*)$ and $(\mathbb{C}^*, \infty)$ are maximal sets of prices leading to arbitrage opportunities for the buyer and the seller of an option, and hence that the interval $[\mathbb{C}_*, \mathbb{C}^*]$ determines the set of *no-arbitrage prices*.

7.4. In the framework of a *three-period* symmetric model of a binomial (B, S)-market consider a European option with contingent claim $f = (S_3 - S_0)^+$. Restricting to self-financing strategies that change only at the time $n = 2$, price this option by reducing to the two-period model of an incomplete market. Prove that $\mathbb{C}^* = S_0 (\lambda^2 - 1)/(\lambda^2 + 1)$ and $\mathbb{C}_* = S_0 (\lambda - 1)/(\lambda + 1)$, where $\lambda > 1$ is the parameter of the original symmetric model.

The eighth lecture introduces the concepts of forward and futures contracts and presents the structure of forward and futures prices under conditions of a complete no-arbitrage market (3.1) (see [1], [31]). A study is made of the simplest model of a forward market with the problem of hedging and optimal investing [5]. The term structure of bond prices and interest rates is also given (for the well-known Ho-Lee model ([21], [30])). The concept of duration is presented.

The structure of prices of other instruments of a financial market. Forwards, futures, bonds

§ 1. Forward and futures contracts: The term structure of forward and futures prices

A *forward contract* is a two-party agreement that must be exercised concerning the purchase or sale of a certain asset (for example, stocks) at a fixed time N in the future for a definite price F.

The price F is called the *strike price*, and N is called the *exercise time* of the contract.

A forward contract, or *forward*, is one of the simple instruments of a financial market that allows one to *insure oneself* against loss and to engage in *speculative activities* on the securities market. The parties taking part in the forward transaction occupy the *long* position (the buyer of the asset) and the *short* position (the seller). The conclusion of the transaction (for example, on the sale of a definite stock) can take place at any time $n = 0, 1, \ldots, N$. Further, one speaks of the opening of the long and short positions, and strike prices F_n, $n = 0, 1, \ldots, N$, that are generally different arise. It is common to call them *forward prices*. It is clear that $F_0 = F$ and $F_N = S_N$.

One of the main problems of "forward" calculations is investigating the structure of the sequence $(F_n)_{n=0,1,\ldots,N}$ and its connection with the prices of the underlying assets B and S of the market (3.1)–(3.2).

At the time a contract is concluded and at the time it is exercised the amounts of the payments are equal, respectively, to zero (entry into the contract does not stipulate any premiums) and $S_N - F_n$, $n = 0, 1, \ldots, N$. Consequently, it is natural to associate a zero contingent claim with such a contract beginning at the time $n = 0, 1, \ldots, N$. As a hedging strategy we consider a G-financing strategy with

$$\Delta G_k^{(n)} = \Delta G_k = \begin{cases} 0, & n \le k < N, \\ S_N - F_n, & k = N. \end{cases}$$

According to the foregoing and Theorem 6.1, the value X of this strategy on a complete (B, S)-market (3.1)–(3.2) is equal to

$$X_n = \mathbf{E}^* \left(\mathcal{E}_n(U) \sum_{k=n+1}^{N} \mathcal{E}_k^{-1}(U) \, \Delta G_k \,\middle|\, \mathcal{F}_n \right)$$
$$= \mathbf{E}^* \left(\mathcal{E}_n (U) \mathcal{E}_{N-1}^{-1}(U) \, (S_N - F_n) \,\middle|\, \mathcal{F}_n \right) = 0,$$

$\mathbf{E}^*$ being the averaging with respect to the martingale measure $\mathbf{P}^*$, $n = 0, 1, \ldots, N$.

Therefore, since $S_n/\mathcal{E}_n$ is a martingale, we get for a deterministic sequence $r_n > -1$ that

(8.1) $$F_n = \mathcal{E}_{N-1}(U) \mathcal{E}_n^{-1}(U) \, \mathbf{E}^* \left(\mathcal{E}_n(U) \mathcal{E}_{N-1}^{-1}(U) \, S_N \,\middle|\, \mathcal{F}_n \right)$$
$$= \mathcal{E}_{N-1}(U) \, \mathbf{E}^* \left((1 + r_N) \mathcal{E}_N^{-1}(U) \, S_N \,\middle|\, \mathcal{F}_n \right)$$
$$= \mathcal{E}_N(U) \mathcal{E}_n^{-1}(U) \, S_n.$$

In particular, it follows from (8.1) that

(8.2) $$F = F_0 = \mathbf{E}^* \left(S_N \,\middle|\, \mathcal{F}_0 \right) = \mathbf{E}^* \, S_N.$$

The relations (8.1) and (8.2) describe the structure of the forward prices F_n in terms of the prices of the underlying assets B and S.

First, forward contracts strive for delivery of the corresponding asset and are *liquid* to a lesser degree than another agreement of forward type called a *futures* contract.

A *futures contract* is an agreement, registered on the stock exchange, about conducting a two-party trading operation with some asset (for example, buying stocks) at a definite time N with the strike price F^* of the asset indicated.

Here the parties need not know each other, since all the repricings are implemented by the *clearing house* of the stock exchange. The last circumstance introduces an essential difference into the structure of the strike prices $(F_n^*)_{n=0,1,\ldots,N}$, which are called the *futures prices*.

In its general features the mechanism of the repricing operation in the clearing house is as follows.

A special account called a *margin* or *margin account* is opened for each party in the contract, and a minimal level is established for it. In opening a position the parties deposit an *initial margin* (usually of the order of 10% of the total cost of the asset in the transaction). At the end of each trading day the position of a client "settles" on the *recomputed price* (the quoted price) determined by the clearing house. The difference between the quoted price and the previous futures price is called the *variational margin*, which is transferred from the loser's account to the winner's account.

Termination of a futures contract means either delivery of the asset, or closing the (long or short) position. More than 90% of futures are closed before the delivery date.

Thus, in the framework of the (B, S)-market (3.1)–(3.2) we have the following sequence of actions in a futures market:

$n = 0 \Longrightarrow$ the parties conclude a futures transaction (on the purchase of a stock) with strike price $F^* = F_0^*$;

$n = i$ $(i = 1, \ldots, N-1) \Longrightarrow$ the clearing house announces a (new) quoted price F_i^* (the new futures price for the stock with respect to this transaction).

If $F_i^* > F_{i-1}^*$, then the seller of the stock loses and deposits the variational margin $F_i^* - F_{i-1}^*$. But if $F_i^* < F_{i-1}^*$, then the buyer must deposit the sum $F_{i-1}^* - F_i^*$. For $n = N$ the buyer pays for the stock at the price F_{N-1}^*, and the seller delivers it.

To study the structure of the futures prices $(F_n^*)_{n=0,1,\ldots,N}$ we employ the method of "backward induction", assuming that *the means of the margin account are invested in bonds* and the level of the margin is equal to $\alpha \in (0, 1)$.

On a complete (B, S)-market we consider *one-step* options with contingent claims

$$f = S_N - F_{N-1}^* + \alpha F_{N-1}^*(1 + r_N), \quad f' = F_{N-1}^* - S_N + \alpha F_{N-1}^*(1 + r_N)$$

which are connected with a futures contract for the purchase and sale, respectively, of the asset S. An amount equal to αF_{N-1}^* is deposited in the margin account.

According to the general theory of option pricing, on a *complete* (B, S)-market (Theorem 4.1) the fair prices of the options introduced are equal, respectively, to

$$\mathbb{C}_{N-1} = \mathbf{E}^* \left(f \, \mathcal{E}_{N-1} \mathcal{E}_N^{-1} \mid \mathcal{F}_{N-1} \right),$$
$$\mathbb{C}_{N-1}' = \mathbf{E}^* \left(f' \, \mathcal{E}_{N-1} \mathcal{E}_N^{-1} \mid \mathcal{F}_{N-1} \right),$$

where, as before, $\mathbf{E}^*$ is the averaging with respect to the unique martingale measure $\mathbf{P}^*$.

Further, using the form of the contingent claim f, we have

(8.3) $\quad \mathbb{C}_{N-1} = \mathbf{E}^* \left(\mathcal{E}_N^{-1} S_N \mid \mathcal{F}_{N-1} \right) \mathcal{E}_{N-1} - (1 + r_N)^{-1} F_{N-1}^* + \alpha F_{N-1}^*$

$$= (1 + r_N)^{-1} \left[\mathbf{E}^* \left(S_N \mid \mathcal{F}_{N-1} \right) - F_{N-1}^* \right] + \alpha F_{N-1}^*.$$

Since the market is arbitrage-free, it follows that $\alpha F_{N-1}^* \geq \mathbb{C}_{N-1}$, and hence, by (8.3),

$$F_{N-1}^* \geq \mathbf{E}^* \left(S_N \mid \mathcal{F}_{N-1} \right).$$

Analogous arguments for the contingent claim f' lead to the relations

$$\mathbb{C}_{N-1}' \geq \alpha F_{N-1}^* \quad \text{and} \quad F_{N-1}^* \leq \mathbf{E}^* \left(S_N \mid \mathcal{F}_{N-1} \right).$$

As a result, since $S/\mathcal{E}$ is a martingale and $\mathcal{E}$ is predictable, we get that

$$F_{N-1}^* = \mathbf{E}^* \left(S_N \mid \mathcal{F}_{N-1} \right) = \mathbf{E}^* \left(S_N \, \mathcal{E}_N^{-1} \mathcal{E}_N \mid \mathcal{F}_{N-1} \right) = (1 + r_N) \, S_{N-1}.$$

Continuing this process, we arrive at the relations

(8.4) $\qquad F_n^* = (1 + r_{n+1}) \cdots (1 + r_N) \, S_n, \qquad n = 0, \ldots, N - 1.$

Here it is obvious that

(8.5) $\qquad F_N^* = S_N \quad \text{and} \quad F_0^* = \mathcal{E}_N S_0 = \mathbf{E}^* S_N.$

The following theorem on the structure of forward and futures prices is the main result of the formulas (8.1)–(8.5).

THEOREM 8.1. *For forward and futures contracts on the same asset in a complete market (3.1)–(3.2) with the deterministic sequence $r_n > -1$ the forward and futures prices satisfy the relations (8.1) and (8.4), respectively, and, in particular, the strike prices F_0 and F_0^* coincide.*

§2. Forward markets and the structure of an optimal investment portfolio

We discuss the question of a forward as an *investment instrument*. Suppose that there is an option on some asset with contingent claim (f, N). In contrast to the preceding, the problem is to find the premium $\mathbb{C}_F(N)$ in realizing the investment activity on a *forward market* when the investment portfolio $\psi_n = (\theta_n, \eta_n)$ consists of θ_n bonds (bank account units) and η_n forward contracts concluded at the time $n - 1$. The corresponding value for $n = 1, \ldots, N$ has the form

$$(8.6) \qquad X_{n-1}^{\psi} = \theta_n B_{n-1},$$

because in concluding a forward contract the parties do not make any payments. All the contracts concluded earlier are exercised only at the time N. Further, we assume that there is neither an influx nor an outflux of capital, and hence its evolution takes place according to the following equations:

$$(8.7) \qquad \Delta X_n^{\psi} = \theta_n \Delta B_n, \qquad n = 1, \ldots, N - 1,$$

$$(8.8) \qquad \Delta X_N^{\psi} = \theta_N \Delta B_N + \sum_{i=1}^{N} \eta_i (S_N - F_{i-1}),$$

where F_i is the forward price at the time i. The class of strategies $\psi = (\psi_n)_{n \geq 1}$ satisfying the relations (8.6), (8.7), and (8.8) is denoted by SF_F. We remark that for the strategies in SF_F the number of bonds in the portfolio is constant and equal to X_0/B_0, while a negative value of η_n is interpreted as the conclusion of $|\eta_n|$ forward contracts on the sale of the asset instead of purchase contracts. Let

$$(8.9) \qquad \gamma_n = \gamma_n(\psi) = \sum_{i=1}^{n} \eta_i.$$

Then, using (8.6)–(8.8) and the equality $F_N = S_N$, we get that

$$\sum_{i=1}^{N} \eta_i (S_N - F_{i-1}) = \sum_{i=1}^{N} \eta_i \sum_{j=i}^{N} (F_j - F_{j-1})$$

$$= \sum_{i=1}^{N} \sum_{j=i}^{N} \eta_i (F_j - F_{j-1}) = \sum_{j=1}^{N} \sum_{i=1}^{j} \eta_i (F_j - F_{j-1}),$$

and hence, by (8.9),

$$(8.10) \qquad X_N^{\psi} = X_0 + \sum_{j=1}^{N} \Delta X_j^{\psi} = X_0 \mathcal{E}_N(U) + \sum_{j=1}^{N} \gamma_j (F_j - F_{j-1}).$$

Using the relation obtained in the preceding section between the forward and basic price of the asset, we have

$$(8.11) \qquad X_N^\psi = \mathcal{E}_N(U) \left[X_0 + \sum_{i=1}^N \gamma_i \left(\mathcal{E}_i^{-1}(U) S_i - \mathcal{E}_{i-1}^{-1}(U) S_{i-1} \right) \right].$$

It follows from (8.11) that the terminal value of the capital corresponding to the strategy ψ coincides with the terminal value of the capital corresponding to the strategy $\pi = \pi(\psi) = (\pi_n)_{n \geq 1}$, $\pi_n = (\beta_n, \gamma_n)$, where β_n is the number of bonds (bank account units), and γ_n is the number of stocks (the underlying asset) in the portfolio of the investor by the time n. The initial value X_0^π is equal to X_0^ψ, and β_n corresponds for all $n = 1, \ldots, N$ to the equation

$$\beta_n = \frac{X_{n-1}^\pi - \gamma_n S_{n-1}}{B_{n-1}}.$$

The strategy π constructed in this way is self-financing.

On the other hand, suppose that π is some strategy in SF, and let

$$(8.12) \qquad \eta_n = \eta_n(\pi) = \gamma_n - \gamma_{n-1}, \quad \theta_n = \frac{X_0^\pi}{B_0}.$$

Then the strategy $\psi = \psi(\pi) = (\psi_n)_{n \geq 1}$, $\psi_n = (\theta_n, \eta_n)$, with $X_0^\psi = X_0^\pi$ belongs to SF_F, and its terminal value coincides with the value of the strategy π.

Thus, we have established a *one-to-one correspondence between the sets* SF_F and SF such that the initial and terminal values of the corresponding strategies coincide.

Next, if $\mathbb{C}(N)$ is the *fair price* of the contingent claim (f, N), and $\pi(f, N) = (\pi_n)_{n \geq 1}$ is a strategy in SF hedging this claim with $X_0^\pi = \mathbb{C}(N)$ and $X_N^\pi = f$ (**P**-a.s.), then there is a strategy $\psi = \psi(\pi) \in \mathrm{SF}_F$ corresponding to π such that $X_0^\psi = \mathbb{C}(N)$ and $X_N^\psi = f$ (**P**-a.s.), and hence $\mathbb{C}_F(N) \leq \mathbb{C}(N)$. Similar arguments show that $\mathbb{C}(N) \leq \mathbb{C}_F(N)$. According to Theorem 4.1,

$$(8.13) \qquad \mathbb{C}_F(N) = \mathbb{C}(N) = \mathbf{E}^* \mathcal{E}_N^{-1}(U) f.$$

Further, according to the formulas (4.5) and (8.12), a hedging strategy $\psi^* = (\psi_n^*)_{n \geq 1}$ is constructed as follows:

$$(8.14) \qquad \theta_n = \frac{\mathbb{C}(N)}{B_0}, \quad \eta_n = \frac{\widetilde{\gamma}_n B_n}{S_{n-1}} - \frac{\widetilde{\gamma}_{n-1} B_{n-1}}{S_{n-2}},$$

where $(\widetilde{\gamma}_n)_{n \geq 1}$ is the predictable sequence in the decomposition (3.12′).

Let us imagine that an investor operating in the complete market (3.1)–(3.2) is trying to attain a value X_N^π as close as possible to some investment goal $f \in L_2(\Omega, \mathcal{F}_N, \mathbf{P})$ by the time N. We denote by SF^2 the set of all strategies $\pi \in \mathrm{SF}$ such that $X_N^\pi \in L^2(\Omega, \mathcal{F}_N, \mathbf{P})$.

Suppose that the investor has initial capital $X_0 = x > 0$ and forms a portfolio $\pi = (\beta, \gamma) \in \mathrm{SF}^2$, so that his capital by the time N is equal to

$$(8.15) \qquad X_N^\pi = \mathcal{E}_N(U) \left[X_0 + \sum_{i=1}^N \gamma_i \mathcal{E}_i^{-1}(U) S_{i-1} (\rho_i - r_i) \right].$$

A portfolio $\widehat{\pi} \in \mathrm{SF}^2$ with initial value x ($\widehat{\pi} \in \mathrm{SF}^2(x)$) is said to be *optimal* if

$$(8.16) \qquad \mathbf{E}\,(X_N^{\widehat{\pi}} - f)^2 = \inf_{\pi \in \mathrm{SF}^2(x)} \mathbf{E}\,(X_N^{\pi} - f)^2.$$

The *optimal investment problem* (8.16) differs from the problem of hedging options (see Lecture 4, § 1) by the presence of only one active party — an investor with fixed initial capital and a fixed investment goal.

Let $Y_N^{\pi} = \mathcal{E}_N^{-1}(U)\,X_N^{\pi} - x$ and $f_N = \mathcal{E}_N^{-1}(U)\,f - x$. It is then clear that for $\widehat{\pi}$ we have the equality

$$(8.17) \qquad \mathbf{E}\,(Y_N^{\widehat{\pi}} - f_N)^2 = \inf_{\pi \in \mathrm{SF}^2(x)} \mathbf{E}\,(Y_N^{\pi} - f_N)^2,$$

which is equivalent to (8.16).

It follows from (8.15) and Theorem 3.1 that Y_n^{π} is a martingale with respect to $\mathbf{P}^*$, where $\mathbf{P}^*$ is a martingale measure (see Lecture 3), and hence

$$(8.18) \qquad \mathbf{E}^*\,Y_n^{\pi} = 0, \qquad n \leq N, \quad \pi \in \mathrm{SF}^2.$$

We introduce the notation

$$\Phi_0 = \{\varphi \in L_2(\Omega, \mathcal{F}_N, \mathbf{P}) : \mathbf{E}^*\,\varphi = 0\},$$

$$z_n = \left.\frac{d\mathbf{P}^*}{d\mathbf{P}}\right|_{\mathcal{F}_n}, \quad \widehat{\varphi} = f_N - z_N \frac{\mathbf{E}^*\,f_N}{\|z_N\|^2},$$

where $\|z_n\|^2 = \mathbf{E}\,z_n^2$.

The indicated variables satisfy the relation

$$(8.19) \qquad \mathbf{E}\,(\widehat{\varphi} - f_N)^2 = \inf_{\varphi \in \Phi_0} \mathbf{E}\,(\varphi - f_N)^2.$$

Indeed, let $\widetilde{\varphi} \in \Phi_0$ be such that $\mathbf{E}\,(\widetilde{\varphi} - f_N)^2 < \mathbf{E}\,(\widehat{\varphi} - f_N)^2$. Then $\widehat{\varphi} = \widetilde{\varphi} - \zeta$, and

$$\mathbf{E}\,(\widetilde{\varphi} - f_N)^2 = \mathbf{E}\left(\zeta - z_N \frac{\mathbf{E}^*\,f_N}{\|z_N\|^2}\right)^2 = \mathbf{E}\,\zeta^2 + (\mathbf{E}^*\,f_N)^2,$$

because

$$\mathbf{E}^*\,\zeta = \mathbf{E}\,z_N \zeta = \mathbf{E}\,(\widetilde{\varphi} z_N - \widehat{\varphi} z_N) = \mathbf{E}^*\,\widetilde{\varphi} - \mathbf{E}^*\,\widehat{\varphi} = 0.$$

However,

$$\mathbf{E}\,(\widehat{\varphi} - f_N)^2 = (\mathbf{E}^*\,f_N)^2 < (\mathbf{E}^*\,f_N)^2 + \mathbf{E}\,\zeta^2 = \mathbf{E}\,(\widetilde{\varphi} - f_N)^2,$$

and we arrive at (8.19).

In view of (8.18) we have that $Y_N^{\pi} \in \Phi_0$ for $\pi \in \mathrm{SF}^2$. On the other hand, suppose that $\varphi \in \Phi_0$. Let us consider the sequence

$$Y_n = \mathbf{E}^*\,(\varphi \,|\, \mathcal{F}_n),$$

which is a martingale with respect to $\mathbf{P}^*$. It follows from the completeness property of the market that

$$(8.20) \qquad Y_n = \sum_{i=1}^{n} \xi_i\,(\rho_i - r_i),$$

where the ξ_i are $\mathcal{F}_{i-1}$-measurable.

Using (8.20), we set

$$
\begin{aligned}
(8.21) \qquad \gamma_i &= \mathcal{E}_i(U)\, S_{i-1}^{-1}\, \xi_i, \\
\beta_i &= [X_{i-1} - \gamma_i\, S_{i-1}]\, \mathcal{E}_{k-1}^{-1}(U),
\end{aligned}
$$

and we get a strategy $\pi = (\beta, \gamma)$ such that $Y_0^\pi = 0$ and $Y_N^\pi = \varphi \in \Phi_0$. Setting $\varphi = \widehat{\varphi}$ and using (8.19), we obtain $\widehat{\pi}$ and arrive at (8.17).

THEOREM 8.2. *Suppose that the investment goal $f \in L_2(\Omega, \mathcal{F}_N, \mathbf{P})$ is specified on the complete (B, S)-market (3.1)–(3.2). Then the formulas (8.22) determine the structure of an optimal investment strategy $\widehat{\pi}$ for (8.16) and (8.17).*

As an illustrative example let us consider the *binomial* model of a (B, S)-market and a deterministic investment goal f. Here the structure of an optimal strategy $\widehat{\pi}$ takes the form

$$
\begin{aligned}
(8.22) \qquad \widehat{\gamma}_n &= \mu\sigma^{-2}\,\mathcal{E}(U)\, S_{n-1}^{-1}\,(f_N - Y_{n-1}^{\widehat{\pi}}), \\
\mu &= \mathbf{E}\,(\rho_n - r_n) = \mathbf{E}\,(\rho_1 - r), \\
\sigma^2 &= \mathbf{E}\,(\rho_n - r_n)^2 = \mathbf{E}\,(\rho_1 - r)^2.
\end{aligned}
$$

It follows from (8.21) and (8.22) that

$$
\begin{aligned}
(8.23) \qquad \widehat{\beta}_n &= \mathcal{E}_{n-1}^{-1}(U) X_{n-1}^{\widehat{\pi}} - \frac{\mu}{\sigma^2}\,\mathcal{E}_n(U)\, S_{n-1}^{-1}\,(f_N - Y_{n-1}^{\widehat{\pi}})\,\frac{S_{n-1}}{\mathcal{E}_{n-1}(U)} \\
&= \mathcal{E}_{n-1}^{-1}\left(X_{n-1}^{\widehat{\pi}} - \mu\sigma^{-2}\left[f - X_{n-1}^{\widehat{\pi}}\,\mathcal{E}_N(U)\mathcal{E}_{n-1}^{-1}(U)\right]\mathcal{E}_n(U)\mathcal{E}_N^{-1}(U)\right), \\
\widehat{\gamma}_n &= \mu\sigma^{-2}\mathcal{E}_n(U)\mathcal{E}_N^{-1}(U)\, S_{n-1}^{-1}\left[f - X_{n-1}^{\widehat{\pi}}\,\mathcal{E}_{n-1}^{-1}(U)\mathcal{E}_N(U)\right].
\end{aligned}
$$

The problem analogous to (8.16) can be regarded also for an investor operating in a *forward contracts market* who is trying to obtain capital X_N^ψ as close as possible to an investment goal $f \in L^2(\Omega, \mathcal{F}_N, \mathbf{P})$ by the time N. Since the basic market is complete and the correspondence between the sets SF and SF_F is one-to-one, the corresponding solution $\widehat{\psi} = (\widehat{\theta}, \widehat{\eta})$ will be given by the formulas

$$
\widehat{\eta}_n = \widehat{\eta}_n(\pi) = \widehat{\gamma}_n - \widehat{\gamma}_{n-1}, \qquad \widehat{\theta}_n = \frac{X_0^{\widehat{\pi}}}{B_0},
$$

where $\widehat{\pi} = (\widehat{\gamma}, \widehat{\beta})$ is an optimal strategy for the problem (8.16).

§3. Bonds: Yield, duration, term structure of prices and interest rates

In the broad sense a *bond* is any promissory note (from a government, a company, and so on). It is issued for a definite period of time N with the obligation to pay its holder an amount A. One speaks of the *exercise (redemption) time N* and the *face value A* of the bond.

During the "lifetime" of a bond it is possible to make intermediate payments called coupon payments, or simply *coupons*. Correspondingly, the bonds become bonds with nonzero and zero coupon. In the latter case it is common to regard the face value to be 1, and the bond itself is called a *bond with discount*.

To begin with, we consider a bond issued on a financial market with bank interest rate $r = \text{const} > 0$. Like every financial instrument, this bond has a price $B(n, N)$ at any time $n \leq N$. It is clear that $B(N, N) = 1$. However, $B(n, N) < 1$

for $n < N$ in view of the fact that if for some $n_0 < N$ the price $B(n_0, N)$ is equal to 1, then it does not make sense to buy such a bond on the given market, because a bank account is more profitable, yielding a return of $(1+r)^{N-n_0} - 1$ by the time N.

Even these simple considerations show the necessity of investigating the *term structure of the prices* $B(n, N)$.

Next, suppose that the coupons $c_1, \ldots, c_N$ are paid on a bond with face value A. We denote the price of such a bond by $B_c(n, N)$.

The *yield* of a bond (at the time $n = 0$) is defined to be the quantity $Y_c = Y_c(0, N)$ determined from the equation

$$(8.24) \qquad B_c(0, N) = \sum_{k=1}^{N} \frac{c_k}{(1+Y_c)^k} + \frac{A}{(1+Y_c)^N}.$$

The coupons are usually paid off in percentages of the face value $c_k = r_c A$, where the quantity r_c is called the *coupon rate*. Therefore, the formula (8.24) can be rewritten in terms of the coupon rate and the face value.

In the case of a bond with a discount, the yield $Y(0, N)$ satisfies the simpler relation

$$(8.25) \qquad B(0, N) = \frac{1}{(1+Y)^N}.$$

The equalities (8.24) and (8.25) show the interrelation between the price of a bond and its yield. It is natural to have this interrelation also for any $n \leq N$. Including the face value A in the coefficient c_N for simplicity, we consider for any $n \leq N$ the quantity $Y_c(n, N)$ such that

$$(8.26) \qquad B_c(n, N) = \sum_{k=n+1}^{N} \frac{c_k}{(1+Y_c(n, N))^{k-n}}.$$

It is then natural to call the quantity $Y_c(n, N)$ in (8.26) the *yield* at the time $n \leq N$.

Further, suppose that on a financial market the interest rate $r(0, N)$ is in effect for a loan given or received at the time 0 over a period $[0, N]$ (*the spot rate*). Then for $B(0, N)$ we have that

$$(8.27) \qquad B(0, N) = \frac{1}{(1+r(0, n))^N}.$$

Comparing (8.25) and (8.27), we see that $Y(0, N) = r(0, N)$. In this connection we say that the yield $Y(0, n)$ represents the *initial structure of the interest rates*.

In the general case we have for a spot rate $r(n, N)$ over the period $[n, N]$:

$$(8.28) \qquad B(n, N) = \frac{1}{(1+r(n, N))^{N-n}}.$$

Consequently, we get from (8.26) and (8.28) the equality $Y(n, N) = r(n, N)$ for any $n \leq N$.

The interrelation among bond prices, yields, and interest rates is clear from the formulas (8.24)–(8.28). It is common to call the whole circle of problems associated with their computation for all $n \leq N$ the *term structure of interest rates*.

A participant in a stock market usually has not one but a whole *portfolio* of bonds. The question arises of whether the concept of yield can be associated with a portfolio. *Duration analysis* provides such a possibility.

We regard the price $B_c(0, Y) > 0$ of a bond with coupon as a function of the yield $Y = Y_c(0, N)$. Assuming that differentiation is possible, we have from (8.24) and (8.26) that

$$(8.29) \qquad \frac{d\,B_c(0, Y)}{dY} = -\sum_{k=1}^{N} \frac{k c_k}{(1+Y)^{k+1}} = \frac{1}{(1+Y)} \sum_{k=1}^{N} \frac{k c_k}{(1+Y)^k} \,.$$

Setting

$$(8.30) \qquad D(Y) = B_c^{-1}(0, Y) \sum_{k=1}^{N} \frac{k c_k}{(1+Y)^k},$$

we get from (8.29) that

$$(8.31) \qquad \frac{d\,B_c(0, Y)}{dY} = -D(Y) \frac{B_c(0, N)}{1+Y} \,.$$

The quantity $D(Y)$ determined by the relation (8.30) is called the *Macaulay duration*. It is clear from (8.29)–(8.31) that the duration shows how much the relative price of a bond changes when the initial yield changes.

Next, we rewrite (8.30) in the form

$$(8.32) \qquad D(Y) = \sum_{k=1}^{N} k\,W_k,$$

where

$$W_k = \frac{c_k\,(1+Y)^{-k}}{\sum_{i=1}^{N} c_i\,(1+Y)^{-i}} \,.$$

We represent a given bond with coupons $c_1, \ldots, c_N$ as a portfolio of zero-coupon bonds with repayment times $1, 2, \ldots, N$. Then, as is clear from (8.32), *the duration is the weighted mean repayment time*. Consequently, duration analysis lets us regard a portfolio of bonds as a single bond with the indicated repayment time.

Let us now consider a bond with discount for which the price dynamics is determined by the recursion relation

$$(8.33) \qquad B(n+1, N) = \frac{B(n, N)}{B(n, n+1)}\, h(\xi_{n+1}; n+1, N),$$

where $\{\xi_n\}_{n \leq N \leq N^*}$ is a sequence of independent random variables taking the values 0 and 1 with probabilities p and $1 - p$, the function $h(i; n, N)$ depends only on the difference $N - n$, $h(\cdot\,; N, N) = 1$, and $h(1; n, N) \geq h(0; n, N)$.

As in the case of the binomial model of the market (3.1), we can take the probability space here to be $\Omega = \{0, 1\}^{N^*}$ and $\mathcal{F}_n = \sigma\{\xi_1, \ldots, \xi_n\}$, with the measure $\mathbf{P}$ completely determined by the "Bernoulli" probability p.

The family $\{B(n, N)\}_{n \leq N^*}$ is said to be *arbitrage-free* if for each $N \leq N^*$ the stochastic sequence

$$\left\{ B_n^{-1} B(n, N) \right\}_{n \leq N}, \qquad B_n^{-1} = \prod_{k=0}^{n} B(k-1, k),$$

is a *martingale* with respect to some measure $\mathbf{P}^*$ that is equivalent to the original measure.

In the case of the model (8.33) of the evolution of bond prices, which is called the *Ho–Lee model*, the no-arbitrage condition leads to the following condition on $p = p_*$ and the *perturbation function h*:

$$p_* h(0; n+1, N) + (1 - p_*) h(1; n+1, N) = 1.$$

Moreover, it can be established directly that there is a $\delta_* > 1$ such that

$$\begin{aligned} h^{-1}(0; n, N) &= p^* + (1 - p^*) \delta_*^{N-n}, \\ h(1; n, N) &= \delta_*^{N-n} \left(p_* + (1 - p_*) \delta_*^{N-n} \right)^{-1}, \end{aligned}$$

(8.34)

and

$$\delta_*^{N-n} = h(1; n, N) h^{-1}(0; n, N).$$

We now define the *forward interest rate* (of a loan received or given over the period $[N, N+1]$) by the relation

$$f(n, N) = \frac{B(n, N)}{B(n, N+1)} - 1,$$

(8.35)

and we set $f(n, n) = r(n)$ as the (instantaneous) spot rate. Note that

$$B(n, N) = \frac{B(0, N)}{B(0, n)} \prod_{j=1}^{n} h^{-1}(\xi_j; j, n) h(\xi_j; j, N).$$

(8.36)

It follows from (8.35) and (8.36) that

$$f(n, N) = \frac{B(0, N)}{B(0, N+1)} \prod_{j=1}^{n} h^{-1}(\xi_j; j, N+1) h(\xi_j; j, N) - 1.$$

(8.37)

Observing now that

$$h(\xi_j; j, N) = \delta_*^{(N-n)\xi_j} h(0; j, N),$$

we get from (8.34) and (8.37) that

(8.38)

$$\begin{aligned} f(n, N) &= \frac{B(0, N)}{B(0, N+1)} \prod_{j=1}^{n} \delta_*^{-\xi_j} \frac{h(0; j, N)}{h(0; j, N+1)} - 1 \\ &= \frac{B(0, N)}{B(0, N+1)} \delta_*^{-(\xi_1 + \cdots + \xi_n)} \frac{p_* + (1 - p_*) \delta_*^N}{p_* + (1 - p_*) \delta_*^{N-n}} - 1 \\ &= \left(\frac{B(0, N)}{B(0, N+1)} - 1 \right) \delta_*^{-(\xi_1 + \cdots + \xi_n)} \frac{p_* + (1 - p_*) \delta_*^N}{p_* + (1 - p_*) \delta_*^{N-n}} \\ &\quad + \delta_*^{-(\xi_1 + \cdots + \xi_n)} \frac{p_* + (1 - p_*) \delta_*^N}{p_* + (1 - p_*) \delta_*^{N-n}} - 1 \\ &= \left[\left(f(0, N) + 1 \right) \delta_*^{-(\xi_1 + \cdots + \xi_n)} \left(p_* + (1 - p_*) \delta_*^N \right) \right. \\ &\quad \left. - \left(p_* + (1 - p_*) \delta_*^{N-n} \right) \right] \left[\left(p_* - (1 + p_*) \delta_*^{N-n} \right]^{-1}. \end{aligned}$$

In particular, we have for the spot rate $r(n)$ that

$$(8.39) \qquad r(n) = \left(f(0,n) + 1\right) \delta_*^{-(\xi_1 + \cdots + \xi_n)} \left(p_* + (1 - p_*)\delta_*^n\right).$$

As a result we get the following theorem about the term structure of bond prices and interest rates in the Ho–Lee model.

THEOREM 8.3. *Suppose that the Ho–Lee model (8.33) is arbitrage-free. Then the formulas (8.36), (8.38), and (8.39) determine the term structure of the prices and interest rates.*

PROBLEMS

8.1. Prove that on a binomial (B,S)-market a forward contract is equivalent to the composition of an option to buy and an option to sell the same asset.

8.2. Prove that for an incomplete (B,S)-market the forward prices F_n, $n = 0,\ldots,N$, and the prices S_n of the underlying asset satisfy the inequalities ($n = 0,\ldots,N$)

$$\inf_{\widetilde{\mathbf{P}} \in \mathbb{P}^*} \widetilde{\mathbf{E}}\left(S_N \mid \mathcal{F}_n\right) \le F_n \le \sup_{\widetilde{\mathbf{P}} \in \mathbb{P}^*} \widetilde{\mathbf{E}}\left(S_N \mid \mathcal{F}_n\right).$$

8.3. On a complete (B,S)-market consider a futures contract with margin level $\alpha_n \in (0,1)$. Assume that the means of the margin account are invested with interest rate $m_n \le r_n$ such that $0 \le \alpha_n(r_n - m_n) < 1$, $n = 1,\ldots,N$. Prove that the following inequalities hold between the forward prices F_n and the futures prices F_n^*, $n = 0,1,\ldots,N$:

$$\frac{F_n}{\Pi_{l=n}^{N}(1 + \alpha_l\,(r_l - m_l))} \le F_n^* \le \frac{F_n}{\Pi_{l=n}^{N}(1 - \alpha_l\,(r_l - m_l))}.$$

8.4. Prove the relations (8.34).

The ninth lecture essentially continues our study of the risk in choosing an investment portfolio. Our approach employs the concept of the yield of the portfolio and the concept of a utility function. The choice of an optimal portfolio is implemented by maximizing its average yield (the "mean–variance" criterion, and so on). See [7], [17], [32], and [34] for the elements of the corresponding theory.

The problem of optimal investment

§ 1. Statement of the investment problem, utility functions

Suppose that a participant in the (B, S)-market (3.1) is given by some positive function $u = u(x)$ and poses the problem

$$(9.1) \qquad \mathbf{E}\, u\!\left(\frac{X_N^\pi}{X_0}\right) \;\to\; \sup,$$

where X_N^π is the value of a self-financing portfolio π with initial value $X_0 > 0$, and the maximum is realized in the class SF.

The functions appearing in the problem (9.1) and determining in many respects its solution are commonly called *utility functions*. This concept is used below only in this narrow sense. For completeness we give the conventional definition of a utility function, and we describe its properties.

A continuously differentiable nondecreasing concave function $u\colon (0, \infty) \to \mathbf{R}^1$ is called a *utility function* if

$$u'(0+) = \lim_{x \downarrow 0} u'(x) = \infty,$$
$$u'(\infty) = \lim_{x \to \infty} u'(x) = 0.$$

We denote by $v = v(x)$ a continuous nonincreasing function inverse to $u'(x)$ such that

$$v\colon (0, \infty) \to (0, \infty),$$
$$v(0+) = \infty, \quad v(\infty) = 0.$$

For $y > 0$ we define the function

$$w(y) = \sup_{x>0} \big[u(x) - xy\big]$$
$$= u(v(y)) - y\,v(y),$$

which is called the *Legendre transform* of the function u.

The function w, being nondecreasing and convex, has the following useful properties:

$$w'(y) = -v(y), \qquad y > 0;$$
$$u(x) = \inf_{y>0} \big[w(y) + xy\big]$$
$$= w(u'(x)) + x\,u'(x), \qquad x > 0.$$

Moreover, for $x, y > 0$ we have the inequalities

$$u(v(y)) \geq u(x) + y\big[v(y) - x\big],$$
$$w(u'(x)) + x(u'(x) - y) \leq w(y).$$

We note also that

$$w(\infty) = u(0+), \quad w(0+) = u(\infty).$$

§ 2. The yield of an investment portfolio and its optimization

Let N be the terminal investment date, $X = (X_n)_{n \leq N}$ a positive stochastic sequence, and η an $\mathcal{F}_N$-measurable nonnegative random variable.

First, we clarify when X is the value of a self-financing portfolio with unit initial and terminal value η.

In a (B, S)-market (3.1) with $r_n \equiv 0$ we consider the class $\mathcal{M}$ of positive martingales Z_n with respect to $\mathbf{P}$ such that $Z_0 = 1$ and $Z_n S_n$ is a martingale with respect to $\mathbf{P}$.

Any element $Z \in \mathcal{M}$ can obviously be thought of as the local density of some *martingale measure* with respect to $\mathbf{P}$, and conversely.

The family $\mathcal{M}$ allows us to characterize the class SF in the following *dual* way: X *is the value of a portfolio* $\pi \in$ SF *if and only if* ZX *is a martingale (with respect to* $\mathbf{P}$) *for all* $Z \in \mathcal{M}$, *and*

(9.2)
$$\eta = \frac{X_N}{X_0} \iff \sup_{Z \in \mathcal{M}} \mathbf{E}\, Z_N \eta \leq 1.$$

We present the solution of the optimization problem (9.1) in correspondence with the foregoing.[1]

Let $Z(\lambda)$ be a solution of the problem

(9.3)
$$\mathbf{E}\, w(\lambda\, Z_N) \to \inf,$$

where the infimum is over the class $\mathcal{M}$.

Let λ^* be a root of the equation

$$\mathbf{E}\, Z(\lambda)\, v(\lambda\, Z(\lambda)) = 1.$$

---

[1] The presentation is based on the approach used by D. O. Kramkov and S. N. Volkov in a study of the same problem for the Black–Scholes model (see Lecture 10).

It is asserted that the maximum in (9.1) is attained for

$$\eta^* = v(\lambda^* Z^*), \quad \text{where} \quad Z^* = Z(\lambda^*),$$

and

$$\mathbf{E}\,u(\eta^*) = \mathbf{E}\,w(\lambda^* Z^*) + \lambda^*.$$

Indeed, if $\eta = Y_N/Y_0$ and Y is the value of $\pi \in \mathrm{SF}$, then $\mathbf{E}\,Z^* \eta \leq 1$. Consequently,

$$\mathbf{E}\,u(y) \leq \mathbf{E}\big[u(\eta) - \lambda^* Z^* \eta\big] + \lambda^*$$
$$\leq \mathbf{E}\,w(\lambda^* Z^*) + \lambda^*,$$

and we need show only that $\eta^* = v(\lambda^* Z^*)$ has the form Y_N/Y_0. In view of (9.2) the latter means that

(9.4) $$\sup_{Z_N \in \mathcal{M}} \mathbf{E}\,Z_N\,\eta^* \leq 1 = \mathbf{E}\,Z^*\,\eta^*.$$

For $Z \in \mathcal{M}$ we consider

$$\psi(\alpha) = \mathbf{E}\,w\big(\lambda^* (\alpha\,Z^* + (1 - \alpha)\,Z)\big).$$

Using the convexity of $\mathcal{M}$, we have that for $\alpha \in [0, 1]$

$$\alpha\,Z^* + (1 - \alpha)\,Z \in \mathcal{M}.$$

Therefore, the function ψ attains its minimum for $\alpha = 1$, and

$$0 \geq \psi'(1) = -\,\mathbf{E}\,v(\lambda^* Z^*)\,(Z^* - Z)$$
$$= \mathbf{E}\,(Z - Z^*)\,\eta^*,$$

which yields (9.4).

Just as the interest rate characterizes the efficiency of the capital deposited in a bank account, the *yield*

$$R(\omega) = \frac{1}{N}\left(\frac{X_N}{X_0} - 1\right), \qquad \omega \in \Omega,$$

of an investment portfolio is an estimate of its efficiency.

The *mean yield* of a portfolio is defined to be $\mathbf{E}\,R$, and the quantity $\sigma = \sqrt{\mathbf{D}\,R}$ is the *variation* of the yield.

The choice of an *optimal* self-financing portfolio starting from the condition of maximal mean yield for a fixed σ is called the *"mean–variance" criterion*:

(9.5) $$\mathbf{E}\,R \;\to\; \max_{\sigma = \text{const}}.$$

We let $r(\sigma)$ be the maximum in (9.5), and we construct the curve $r(\sigma)$ by giving the parametric equations

$$r = r(c)$$

and

$$\sigma = \sigma(c), \qquad c > 0.$$

To solve the extremal problem (9.5) we introduce the "utility function"

$$u(x) = x - c\frac{x^2}{2}, \qquad c > 0.$$

It is clear that

$$v(x) = c^{-1}(1 - x),$$
$$w(x) = (2c)^{-1}(1 - x)^2.$$

The use of the approach (9.3)–(9.4) presented above gives the following expressions for r and σ:

$$r(c) = \frac{1}{Nc}\left(1 - c - \lambda(c)\,\mathbf{E}\,Z(c)\right),$$
$$\sigma(c) = \frac{\lambda(c)}{Nc}\sqrt{\mathbf{E}\left(Z(c)\right)^2 - \left(\mathbf{E}\,Z(c)\right)^2},$$

where $Z(c) \in \mathcal{M}$ and $\lambda(c)$ are determined by the relations

$$Z(c) = \arg\min \mathbf{E}\left(1 - \lambda(c)\,Z(c)\right)^2,$$
$$\mathbf{E}\,Z(c)\left(1 - \lambda(c)\,Z(c)\right) = c.$$

Then the terminal value of the optimal strategy has the form

$$Y_N(c) = Y_0(c)\,\frac{1 - \lambda(c)\,Z(c)}{c}.$$

If we now introduce the *yield* of the portfolio

$$R(\beta) = R(\omega, \beta)$$
$$= \frac{1}{\beta N}\left(\left(\frac{X_N}{X_0}\right)^\beta - 1\right),$$

which is an analogue of the interest rate $(\beta > 0)$ and the discounted rate $(\beta < 0)$, then again we can solve the problem (9.5).

Denoting the corresponding maximum by $r = r(\beta)$, we consider here the different utility function

$$u(x) = \beta^{-1}x^\beta.$$

Here the function v and the Legendre transform are determined by the formulas

$$v(x) = x^{1/(\beta-1)},$$
$$w(x) = \frac{1 - \beta}{\beta}\,x^{\beta/(\beta-1)}.$$

In a way similar to the preceding we have that

$$r(\beta) = \frac{1}{N\beta}\left(\left(\mathbf{E}\,\widehat{Z}^{\beta/(\beta-1)}\right)^{1-\beta} - 1\right),$$

where

$$\widehat{Z} = \arg\min\left[\frac{1 - \beta}{\beta}\,\mathbf{E}\,Z^{\beta/(\beta-1)}\right].$$

Here the terminal value of the optimal strategy has the form

$$Y_N(\beta) = Y_0(\beta)\,\frac{\widehat{Z}^{\,1/(\beta-1)}}{\mathbf{E}\,\widehat{Z}^{\,\beta/(\beta-1)}}.$$

§ 3. Optimal investment in the binomial model of a (B, S)-market

In this section we study the following form of the problem (9.1) for a binomial (B, S)-market (3.1) with the condition (4.14): find a strategy $\pi^* \in \mathrm{SF}$ with value $X_n^*(x) = X_n^{\pi^*} = X_n^{\pi^*}(x)$, $X_0^* = x$, such that

$$(9.6) \qquad \mathbf{E}u(X_N^*(x)) = \sup_{\pi \in \mathrm{SF}} \mathbf{E}u(X_N^\pi(x)) = \varphi(x),$$

where u is a given utility function.

Assuming, without loss of generality, that the interest rate r is zero, we denote by Z_n^* the local density of the unique martingale measure $\mathbf{P}^*$ with respect to $\mathbf{P}$. We consider the function $\psi(y) = \mathbf{E}w(yZ_N^*)$, $y > 0$, and we prove that

$$(9.7) \qquad \varphi(x) = \inf_{y>0} (\psi(y) + xy), \quad x > 0.$$

It follows from the definition of the Legendre transform that for any $\pi \in \mathrm{SF}$

$$w(yZ_N^*) \geq u(X_N^\pi(x)) - yZ_N^* X_N^\pi(x),$$
$$\mathbf{E}w(yZ_N^*) \geq \mathbf{E}u(X_N^\pi(x)) - y\mathbf{E}^* X_N^\pi(x)$$
$$= \mathbf{E}u(X_N^\pi(x)) - yx,$$

and hence for $x > 0$ and $y > 0$

$$(9.8) \qquad \varphi(x) \leq \psi(y) + xy.$$

Further, by Theorem 4.1 there exists for any contingent claim (f, N) a *minimal hedge* with initial value $x_f = \mathbf{E}^* f$, and therefore

$$(9.9) \qquad \mathbf{E}u(f) \leq \varphi(x_f).$$

It follows from (9.8) and (9.9) that

$$\psi(y) = \sup_{f>0} \mathbf{E}(u(f) - yZ_N^* f)$$
$$\leq \sup_{x:\mathbf{E}^* f=x} \mathbf{E}(u(f) - yZ_N^* f)$$
$$\leq \sup_{x>0}(\varphi(x) - xy),$$

and this implies (9.7).

We now establish the key result in this section.

THEOREM 9.1. *Let $x > 0$ and $\widehat{y} > 0$ be such that $\varphi(x) = \psi(\widehat{y}) + x\widehat{y}$. Then the terminal value $X_N^*(x)$ of an optimal strategy $\pi^* \in \mathrm{SF}$ for the problem (9.6) is determined by the formula*

$$(9.10) \qquad X_N^*(x) = X_N^{\pi^*}(x) = v(\widehat{y}Z_N^*),$$

where v is the function inverse to u'.

PROOF. For any $y > 0$ the terminal value of an optimal portfolio is equal to an $\mathcal{F}_N$-measurable random variable f such that

$$(9.11) \qquad u(f) - yZ_N^* f = \sup_{x>0}(u(x) - yZ_N^* x) = w(yZ_N^*),$$

and therefore $u'(f) = yZ_N^*$.

For the $\widehat{y}$ in the condition of the theorem we denote the terminal value by $\widehat{f}$, and we show that $\varphi(x) = \mathbf{E}u(\widehat{f})$. By virtue of (9.11),

$$\mathbf{E}u(\widehat{f}) - \mathbf{E}(\widehat{y}Z_N^* \widehat{f}) = \mathbf{E}w(\widehat{y}Z_N^*) = \psi(\widehat{y}).$$

Therefore, the inequality $\mathbf{E}Z_N^* \widehat{f} \geq x$ gives us that

$$\mathbf{E}u(\widehat{f}) \geq \psi(\widehat{y}) + x\widehat{y}.$$

On the other hand, it follows from the condition of the theorem that

$$\mathbf{E}u(\widehat{f}) \geq \sup_{\pi \in \mathrm{SF}} \mathbf{E}u(X_N^\pi(x)).$$

Consequently,

$$\mathbf{E}u(\widehat{f}) = \sup_{\pi \in \mathrm{SF}} \mathbf{E}u(X_N^\pi(x)) = \varphi(x),$$

which finishes the proof.

As is clear from the preceding, we must find the density Z_N^*. Since $(Z_n^*)_{n \leq N}$ is a martingale, Lemma 4.1 gives us that

$$Z_N^* = 1 + \sum_{k=1}^{N} \varphi_k(\rho_k - m),$$

where $(\varphi_k)_{k \leq N}$ is a predictable sequence, and $m = \mathbf{E}\rho_1$. Starting from this representation, we write a more convenient (for what follows) *multiplicative* decomposition for Z_N^* (see (2.5')):

$$(9.12) \qquad Z_N^* = \mathcal{E}_N(\widehat{Z}^*),$$

where $\Delta \widehat{Z}_k^* = (Z_{k-1}^*)^{-1}\varphi_k(\rho_k - m) \equiv \psi_k(\rho_k - m)$.

It follows from the equality $r = 0$ and the martingale property of the measure $\mathbf{P}^*$ that for $n \leq N$

$$\mathbf{E}(Z_n^* \rho_n \mid \mathcal{F}_{n-1}) = 0,$$
$$\mathbf{E}((1 + \psi_n(\rho_n - m))\rho_n \mid \mathcal{F}_{n-1}) = 0.$$

These relations lead to an expression of the coefficients ψ_n of the multiplicative decomposition in terms of the mean $m = \mathbf{E}\rho_1$ and the variance $\sigma^2 = \mathbf{D}\rho_1$:

$$\mathbf{E}\rho_n + \mathbf{D}\rho_n \psi_n = 0 \implies \psi_n = -\frac{m}{\sigma^2},$$

and (9.12) is transformed to the form

$$(9.13) \qquad Z_N^* = \mathcal{E}_N\left(\sum (-m/\sigma^2)(\rho_k - m)\right).$$

Let us now consider the logarithm utility function $u(x) = \ln x$; in this case $v(x) = 1/x$. We must find $\varphi(x) = \sup_{\pi \in \mathrm{SF}} \mathbf{E}\ln(X_N^\pi(x))$, the terminal value $X_N^*(x)$, for which $\varphi(x) = \mathbf{E}\ln X_N^*(x)$, and a strategy $\pi^* \in \mathrm{SF}$ with this terminal value.

In view of the form $X_N^* = 1/(yZ_N^*)$ of the terminal value the function $\psi(y) = \mathbf{E}\sup_{X_N > 0}(\ln X_N - yZ_N^* X_N)$ can be rewritten as

$$\psi(y) = \mathbf{E}\left(\ln \frac{1}{yZ_N^*} - yZ_N^* \cdot \frac{1}{yZ_N^*}\right) = -\ln y - 1 - \mathbf{E}\ln Z_N^*,$$

and hence

$$\varphi(x) = \inf_{y > 0}(\psi(y) + yx) = \inf_{y > 0}(-\ln y + yx) - 1 - \mathbf{E}\ln Z_N^*.$$

Since $\hat{y} = 1/x$ minimizes $-\ln y + yx$, the cost function is

(9.14) $$\varphi(x) = \ln x - \mathbf{E}\ln Z_N^*.$$

Here the terminal value is determined from (9.10):

(9.15) $$X_N^*(x) = \frac{1}{yZ_N^*} = \frac{x}{Z_N^*}.$$

To find the optimal strategy $\pi_n^* = (\beta_n^*, \gamma_n^*)$ it is convenient, as in §7.3, to introduce the *proportion* $\alpha_n^* = \gamma_n S_{n-1}/X_{n-1}^*$ of risky capital in the portfolio; this is a predictable sequence. By (7.19),

$$X_N^* = x \cdot \mathcal{E}_N\left(\sum \alpha_k^* \rho_k\right).$$

From this we get by using (9.13) and (9.15) that

(9.16) $$\mathcal{E}_N\left(\sum \alpha_k^* \rho_k\right)\mathcal{E}_N\left(\sum\left(-\frac{m}{\sigma^2}(\rho_k - m)\right)\right) = 1.$$

We show that (9.16) implies that for $n = 1, \dots, N$

(9.17) $$\alpha_n^* = -\frac{m}{ab}.$$

From induction arguments it suffices to verify (9.17) in the case $N = 1$. In this case (9.16) can be rewritten in the form

$$(1 - \alpha_1^* \rho_1)\left(1 - \frac{m}{\sigma^2}(\rho_1 - m)\right) = 1,$$

and on the set $\{\omega : \rho_1 = b\}$ this reduces to

$$(1 + \alpha_1^* b)\left(1 - \frac{m}{\sigma^2}(b - m)\right) = 1.$$

The solution of the latter is $\alpha_1^* = -m/(ab)$. We arrive at precisely the same result on the set $\{\omega : \rho_1 = a\}$.

Using (9.17), we obtain formulas for the *optimal strategy*:

(9.18) $$\gamma_n^* = -\frac{mX_{n-1}^*}{abS_{n-1}}, \qquad \beta_n^* = X_{n-1}^* + \frac{mX_{n-1}^*}{ab}.$$

This proves the following result.

THEOREM 9.2. *In the optimal investment problem* (9.6) *with logarithm utility function* $u(x) = \ln x$ *the function* $\varphi(x)$, *the optimal terminal value* X_N^*, *and the optimal strategy* $\pi^* = (\beta_n^*, \gamma_n^*)$ *are given by the formulas* (9.14), (9.15), *and* (9.18).

PROBLEMS

9.1. For the single-period binomial model of a (B, S)-market, maximize the mean yield $R = R^\pi$ in the class $\mathrm{SF}(\delta)$ of strategies $\pi_n = (\beta_n, \gamma_n)$ such that

$$B_{n-1} \Delta \beta_n + S_{n-1} \Delta \gamma_n = -\delta \, S_{n-1} \, |\Delta \gamma_n|, \qquad \delta > 0.$$

9.2. For the single-period binomial model of a (B, S)-market, solve the following investment problem:

$$\mathbf{E} \, (X_1^\pi - f)^2 \to \inf_{\mathrm{SF}(\delta)},$$

where $X_0^\pi = x$ and f is the deterministic goal (cf. (8.16)).

9.3. Solve the optimization problem (9.6) for the power utility function $u(x) = x^\alpha / \alpha$, $\alpha \in (-\infty, 1)$, $\alpha \neq 0$, using the method in §3.

In the tenth and last lecture we discuss the question of continuous models of a (B, S)-market. First, we show that any discrete model can be looked at on a "continuous" time interval, where prices change like random step-processes. Second, for the binomial model we derive "limiting" (as the discreteness step Δ goes to 0) continuous models (Black–Scholes [15] and Merton [35], [36]). In this way we prove that the famous Black–Scholes formula is a limiting form of the Cox–Ross–Rubinstein formula ([18], [19]; see also [8], [21]).

The concept of continuous models.
Limiting transitions from a discrete market
to a continuous one. The Black–Scholes formula

§1. (B, S, Δ)-markets and continuous models

The discreteness of the market (3.1) was determined by the *unit* time step of the change in the prices of bonds and stocks. However, with some generally nonintegral moment of time $T > 0$ specified, it is not hard to imagine that the evolution of prices takes place with a step that is a multiple not of 1 but of some positive number Δ. As a result, for $t = \Delta, 2\Delta, \ldots$ we arrive at the following recursion equations determining a (B, S, Δ)-*market*:

$$
(10.1) \qquad
\begin{aligned}
B_t^{(\Delta)} &= \left(1 + r_t(\Delta)\right) B_{t-\Delta}^{(\Delta)}, & B_0^{(\Delta)} &> 0, \\
S_t^{(\Delta)} &= \left(1 + \rho_t(\Delta)\right) S_{t-\Delta}^{(\Delta)}, & S_0^{(\Delta)} &> 0,
\end{aligned}
$$

where $(r_t(\Delta))$ and $(\rho_t(\Delta))$ are stochastic sequences that are measurable with respect to $\mathcal{F}_t^{(\Delta)}$, $t = 0, \Delta, \ldots, [T/\Delta]\Delta$.

In particular, for $\Delta = 1$ and $[T/\Delta] = N$ we get the discrete (B, S)-market (3.1).

We extend the (B, S, Δ)-market to *the whole interval* $[0, T]$ as follows: for $s \in [t, t + \Delta)$, $t = 0, \Delta, 2\Delta, \ldots$

$$
(10.2) \qquad
\begin{aligned}
B_s^{(\Delta)} &\equiv B_t^{(\Delta)}, & S_s^{(\Delta)} &\equiv S_t^{(\Delta)}, & \mathcal{F}_s^{(\Delta)} &\equiv \mathcal{F}_t^{(\Delta)}, \\
r_s(\Delta) &\equiv r_t(\Delta), & \rho_s(\Delta) &\equiv \rho_t(\Delta).
\end{aligned}
$$

In this case we speak of $B_t^{(\Delta)}$, $S_t^{(\Delta)}$, $r_t(\Delta)$, and $\rho_t(\Delta)$ as *random processes* defined on the interval $[0, T]$. Here the differences of the form $B_{t_2}^{(\Delta)} - B_{t_1}^{(\Delta)}, \ldots$ are called the *increments* of these random processes on the interval $[t_1, t_2] \subseteq [0, T]$.

Further, if on the (B, S, Δ)-market (10.1) we specify the contingent claim

$$f = \left(S_{[T/\Delta]\,\Delta} - K\right)^{+},$$

then we get a European call option.

The natural question arises as to the behavior of the premium $\mathbb{C}_T(\Delta)$ for this option since $\Delta \to 0$. It is clear from (10.1) that the given *asymptotic problem* presupposes the existence of a whole family (with respect to $\Delta > 0$) of models (10.1) specified on the interval $[0, T]$ being refined.

A *continuous model* can be thought of as a limit of such discrete models. Here the limits of the random step processes $B_t^{(\Delta)}, S_t^{(\Delta)}, \ldots$ will also be random processes which are either continuous or right-continuous as functions of t.

In order to form an idea of the possible "purely" continuous models, we consider the case when $r_t(\Delta) \equiv r\,\Delta$ $(r > 0)$ and $\rho_t(\Delta)$ are independent random variables $(t = \Delta, 2\Delta, \ldots)$ distributed like the Bernoulli random variable $\rho(\Delta)$ with values $a(\Delta) < b(\Delta)$.

We now use (10.2) to represent the variable $\rho_t(\Delta)$, $t \in [0, T]$, in the form

$$\rho_t(\Delta) = \mu\,\Delta + \sigma\,\Delta M_t,$$

where $\mathbf{E}\,\rho_t(\Delta) = \mu\,\Delta$, $\mathbf{D}\,\Delta M_t = \Delta$, $\mathbf{D}\,\rho_t(\Delta) = \sigma^2\Delta$.

As a result we get the following representation of the model (10.1), which is defined for all $t \in [0, T]$:

(10.3)
$$\Delta B_t^{(\Delta)} = r\,B_{t-}^{(\Delta)}\,\Delta,$$
$$\Delta S_t^{(\Delta)} = \left(\mu\,\Delta + \sigma\,\Delta M_t\right)S_{t-}^{(\Delta)},$$

where the "limits from the left" $B_{t-}^{(\Delta)}$ and $S_{t-}^{(\Delta)}$ are equal to $B_{t-\Delta}^{(\Delta)}$ and $S_{t-\Delta}^{(\Delta)}$, while the random process M_t with zero mean has independent increments ΔM on disjoint time intervals.

In view of (10.3) it is natural to expect that the limiting model must be the following:

(10.4)
$$dB_t = r B_{t-}\,dt,$$
$$dS_t = \left(\mu\,dt + \sigma\,dM_t\right)S_{t-},$$

where the *differentials* dB_t, dS_t, and dM_t of the random processes are understood solely as certain *formal* limits of the increments ΔB_t, ΔS_t, and ΔM_t as $\Delta \to 0$.

Further, the limits of Bernoullian random variables are well known to be Gaussian and Poisson variables. Therefore, the following two types of limit models (10.4) are possible.

The *Black–Scholes model*: $M_t = W_t$ is a Wiener process (Brownian motion) having continuous sample paths $W_{\centerdot}(\omega)$, $\omega \in \Omega$, and independent Gaussian increments, and satisfying $\mathbf{E}\,W_t = 0$ and $\mathbf{D}\,W_t = t$.

The *Merton model*: $M_t = \Pi_t - \lambda t$ is a "centered Poisson process" Π_t with intensity λ.

The parameters r, μ, and σ of the model (10.4) are called, respectively, the *interest rate*, the *appreciation rate*, and the *volatility* of the market.

Just as in the discrete case, we introduce for (10.4) the concept of a European call option with contingent claim $(S_T - K)^+$ and its fair price $\mathbb{C}_T$.

It is natural to expect that under certain conditions

(10.5) $$\mathbb{C}_T(\Delta) \to \mathbb{C}_T \quad \text{as} \quad \Delta \to 0.$$

These conditions are discussed in the next section.

§2. The Black–Scholes formula

Suppose that the parameters of the (B, S, Δ)-market (10.1) satisfy the relations

(10.6) $$1 + r(\Delta) = e^{r\Delta}, \quad 1 + b(\Delta) = e^{\sigma\sqrt{\Delta}}, \quad 1 + a(\Delta) = e^{-\sigma\sqrt{\Delta}}$$

with positive constants r and σ.

According to the Cox–Ross–Rubinstein formula,

(10.7) $$\mathbb{C}_T(\Delta) = S_0 \, \mathbb{B}\big(k_0(\Delta), [T/\Delta], \widetilde{p}_\Delta\big)$$
$$- K\big(1 + r(\Delta)\big)^{-[T/\Delta]} \, \mathbb{B}\big(k_0(\Delta), [T/\Delta], p_\Delta^*\big),$$

where

$$k_0(\Delta) = 1 + \left[\frac{\ln\left(K/(S_0(1+a)^{[T/\Delta]})\right)}{\ln\left((1+a)/(1+b)\right)}\right],$$

$$p_\Delta^* = \frac{r(\Delta) - a(\Delta)}{b(\Delta) - a(\Delta)},$$

$$\widetilde{p}_\Delta = \frac{1 + b(\Delta)}{1 + a(\Delta)} p_\Delta^*.$$

Using the de Moivre–Laplace theorem for $\Delta \to 0$, we get that

(10.8)
$$\mathbb{B}\big(k_0(\Delta), [T/\Delta], p_\Delta^*\big) \sim \Phi\left(\frac{[T/\Delta]\, p_\Delta^* - k_0(\Delta)}{\sqrt{[T/\Delta]\, p_\Delta^*\, (1 - p_\Delta^*)}}\right) = \Phi(y_\Delta^*),$$

$$\mathbb{B}\big(k_0(\Delta), [T/\Delta], \widetilde{p}_\Delta\big) \sim \Phi\left(\frac{[T/\Delta]\, \widetilde{p}_\Delta - k_0(\Delta)}{\sqrt{[T/\Delta]\, \widetilde{p}_\Delta\, (1 - \widetilde{p}_\Delta)}}\right) = \Phi(\widetilde{y}_\Delta),$$

where

$$\Phi(x) = \frac{1}{\sqrt{2\pi}} \int_{-\infty}^{x} e^{-t^2/2} dt.$$

It follows from (10.7) and (10.8) that as $\Delta \to 0$

(10.9) $$\mathbb{C}_T(\Delta) \sim S_0 \, \Phi(\widetilde{y}_\Delta) - K\big(1 + r(\Delta)\big)^{-[T/\Delta]} \, \Phi(y_\Delta^*).$$

Consequently, to prove (10.5) it remains to establish the asymptotic behavior of the quantities $k_0(\Delta)$, $(1 + r(\Delta))^{-[T/\Delta]}$, $\widetilde{y}_\Delta$, and y_Δ^* as $\Delta \to 0$.

It is clear that as $\Delta \to 0$

$$k_0(\Delta) \sim \frac{\ln\left(K/S_0\right) + [T/\Delta]\, \sigma\sqrt{\Delta}}{2\sigma\sqrt{\Delta}},$$

$$\big(1 + r(\Delta)\big)^{-[T/\Delta]} \sim e^{-rT}.$$

Further, it can be established directly that as $\Delta \to 0$

$$[T/\Delta]\,p_\Delta^* = \frac{T\Delta\,(r - \sigma^2/2) + T\sigma\sqrt{\Delta} + o\,(\Delta)}{2\sigma\Delta^{3/2} + o\,(\Delta^{3/2})},$$

$$[T/\Delta]\,p_\Delta^* - k_0 = \frac{T\Delta\,(r - \sigma^2/2) + \Delta\ln\,(S_0/K) + o\,(\Delta)}{2\sigma\Delta^{3/2} + o\,(\Delta^{3/2})},$$

$$\sqrt{[T/\Delta]\,p_\Delta^*\,(1 - p_\Delta^*)} = \sqrt{\frac{T}{\Delta}}\,\sqrt{\frac{\sigma^2\Delta + o\,(\Delta)}{4\sigma^2\Delta + o\,(\Delta)}}.$$

Consequently,

$$\lim_{\Delta \to 0} \frac{[T/\Delta]\,p_\Delta^* - k_0(\Delta)}{\sqrt{[T/\Delta]\,p_\Delta^*\,(1 - p_\Delta^*)}} = \frac{T\,(r - \sigma^2/2) + \ln\,(S_0/K)}{\sigma\sqrt{T}} = y^*.$$

It can be established similarly that

$$\lim_{\Delta \to 0} \frac{[T/\Delta]\,\widetilde{p}_\Delta - k_0(\Delta)}{\sqrt{[T/\Delta]\,\widetilde{p}_\Delta\,(1 - \widetilde{p}_\Delta)}} = \frac{T\,(r + \sigma^2/2) + \ln\,(S_0/K)}{\sigma\sqrt{T}} = \widetilde{y}.$$

As a result, (10.9) gives us the *Black–Scholes formula*

(10.10) $$\mathbb{C}_T(\Delta) \to \mathbb{C}_T = S_0\,\Phi(\widetilde{y}) - K\,e^{-rT}\,\Phi(y^*).$$

The formula (10.10) establishes a *fair price* for a European call option in the framework of the continuous Black–Scholes model (10.4).

For completeness we also present briefly the Poisson asymptotics in (10.5) under the assumption that

(10.11) $$1 + r(\Delta) = e^{r\Delta}, \quad 1 + a(\Delta) = e^{-\sigma\Delta}, \quad 1 + b(\Delta) = 1 + b.$$

It follows from (10.11) that as $\Delta \to 0$

$$p_\Delta^* = \frac{r + \sigma}{b}\,\Delta + o\,(\Delta),$$

$$\widetilde{p}_\Delta = \frac{1 + b}{b}\,(r + \sigma)\,\Delta + o\,(\Delta),$$

$$k_0(\Delta) \sim \left[-\frac{\ln\,(K/S_0) + \sigma T}{\ln\,(1 + b)} \right] = k_0.$$

It is clear that as $\Delta \to 0$

$$[T/\Delta]\,p_\Delta^* \to \frac{r + \sigma}{b}\,T, \quad [T/\Delta]\,\widetilde{p}_\Delta \to \frac{1 + b}{b}\,(r + \sigma)\,T,$$

and hence by Poisson's theorem,

$$\mathbb{B}(k_0(\Delta), [T/\Delta], \widetilde{p}_\Delta) \longrightarrow P_1 = \sum_{i=k_0}^{\infty} \frac{[(1 + b)(r + \sigma)T/b]^i}{i\,!}\,\exp\{-(1 + b)(r + \sigma)T/b\},$$

$$\mathbb{B}(k_0(\Delta), [T/\Delta], p_\Delta^*) \longrightarrow P_2 = \sum_{i=k_0}^{\infty} \frac{[(r + \sigma)T/b]^i}{i\,!}\,\exp\{-(r + \sigma)T/b\}$$

as $\Delta \to 0$.

As a result, we get from (10.5) and (10.6) that

(10.12) $$\mathbb{C}_T(\Delta) \to \mathbb{C}_T = S_0 P_1 - K e^{-rT} P_2$$

as $\Delta \to 0$.

The formula (10.12) gives an idea of how to compute a *fair price* of a European call option in the framework of the Merton model (10.4).

PROBLEMS

10.1. Let $\mathbb{C} = \mathbb{C}(S_0, T, K, \sigma, r)$ be the fair price of a call option in the Black–Scholes model, where S_0 is the initial price of a stock, T exercise time, K is the exercise price, σ is the volatility, and r is the interest rate. Prove that

a) $\lim_{S_0 \to 0} \mathbb{C}(S_0) = 0$, $\lim_{S_0 \to \infty} \mathbb{C}(S_0) = \infty$;

b) $\lim_{K \to 0} \mathbb{C}(K) = S_0$, $\lim_{k \to \infty} \mathbb{C}(K) = \infty$;

c) $\lim_{T \to \infty} \mathbb{C}(S_0, T, K) = 0$ if $S_0 < K$;
   $\lim_{T \to 0} \mathbb{C}(S_0, T, K) = S_0 - K$ if $S_0 > K$;
   $\lim_{T \to \infty} \mathbb{C}(S_0, T, K) = S_0$ if $S_0 > K$;

d) $\lim_{T \to \infty} \mathbb{C}(S_0, T, K, \sigma, r) = 0$ if $S_0 < Ke^{-rT}$;
   $\lim_{\sigma \to 0} \mathbb{C}(S_0, T, K, \sigma, r) = S_0 - Ke^{-rT}$ if $S_0 > Ke^{-rT}$;
   $\lim_{\sigma \to \infty} \mathbb{C}(S_0, T, K, \sigma, r) = S_0$ if $S_0 > Ke^{-rT}$;

e) $\lim_{r \to \infty} \mathbb{C}(r) = S_0$.

10.2. Suppose that the Black–Scholes price for a standard European call option is a smooth function $\mathbb{C}(t, x)$. Passing to the limit from the binomial model to a continuous model (with interest rate r and volatility σ), prove that $\mathbb{C}$ satisfies the following equation of Black and Scholes:

$$\frac{1}{2}\sigma^2 x^2 \frac{\partial^2 \mathbb{C}}{\partial x^2} + rx\frac{\partial \mathbb{C}}{\partial x} + \frac{\partial \mathbb{C}}{\partial t} - r\,\mathbb{C} = 0.$$

Appendix 1

We consider the following model of a *discrete financial market*. Let B_n be the price of a bond at the time $n = 0, 1, \ldots$, and suppose that

$$B_{n+1} = B_n (1 + r), \qquad B_0 = 1, \quad n \leq N - 1.$$

Let S_n be the price of a stock at the time n, and suppose that it is determined from the equation

$$\frac{S_{n+1}}{B_{n+1}} = \frac{S_n}{B_n} + \rho_n, \qquad S_0 > 0, \quad n \leq N - 1,$$

where r is the interest rate, and $\{\rho_n\}$ is a sequence of independent identically distributed random variables taking the values $\pm \eta$, $\eta > 0$.

Assuming there is a spread between the ask and bid prices of the shares, there are dividends, and there is a payment for leasing (borrowing) the stock, we take these quantities to be equal to δB_{n+1}, αB_{n+1}, and θB_{n+1}, respectively.

Let $\mathbb{C}_*$ be the lower (bid) price and $\mathbb{C}^*$ the upper (ask) price of a European call option with contingent claim $f = (S_T - K)^+$. Also:

σ is the volatility (changeability) of the stock price;

R^* and R_* are the borrowing rate and the lending rate;

P is the payment coefficient for leasing a single share;

Q is the dividend payment coefficient on a single share.

We specify the parameters of the model:

N is the number of layers of the binomial lattice into which $[0, T]$ is divided;

$T = 30 \, \text{days}/360 \, \text{days} = 0.08$;

x_0 is the current stock price, which equals 75 (rubles) initially;

$K = 75$ (rubles), the redemption price;

$r^* = R^* T/N$, the borrowing rate;

$r_* = R_* T/N$, the lending rate;

$$\alpha = Q\frac{T}{N}, \quad \theta = P\frac{T}{N}, \quad \eta = \sigma\sqrt{\frac{T}{N}}, \quad \sigma = 8 \text{ (rubles)};$$
$$R^* = 0.15, \quad R_* = 0.08, \quad P = 0.04;$$
$$Q = 0.04, \quad \delta = 0.04, \quad N = 4.5.$$

These parameters were used in an actual situation involving quotation of options on the Russian market in 1995. We present the computational formulas obtained by the "backward induction" method.

1. For the ask price of a call option we have:

$$p_1 = \frac{\eta + \delta - \alpha}{2\eta + \delta}, \qquad q_1 = \frac{\eta - \alpha}{2\eta + \delta},$$

$$f(N, x) = \left(x\,(1 + r^*)^N - K\right)^+,$$

$$f(n, x + \delta/2) = \frac{p_1\, f(n + 1, x + \eta + \delta/2) + (1 - p_1)\, f(n + 1, x - \eta - \delta/2)}{1 + r^*},$$

$$f(n, x - \delta/2) = \frac{q_1\, f(n + 1, x + \eta + \delta/2) + (1 - q_1)\, f(n + 1, x - \eta - \delta/2)}{1 + r^*},$$

where $x = x_0 \pm i\eta$, $i = 0, 1, \ldots, n$, and here

$$\mathbb{C}^* = f(0, x_0 + \delta/2)$$

$$\gamma_0 = \frac{f(1, x_0 + \eta + \delta/2) - (1 + r^*)\, f(0, x_0 + \delta/2)}{(\eta + \alpha)(1 + r^*)},$$

$$\beta_0 = f(0, x_0 + \delta/2) - \gamma_0(x_0 + \delta/2)$$

are, respectively, the ask price, number of shares, and number of bonds in the portfolio given as the answer.

2. For the bid price of the call option we have:

$$p_2 = \frac{-\alpha - \theta + \eta}{2\eta - \delta}, \qquad q_2 := \frac{-\alpha - \theta + \eta - \delta}{2\eta - \delta},$$

$$f(N, x) = G_N(x) = \left(x\,(1 + r_*)^N - K\right)^+$$

$$f(n, x + \delta/2) = \frac{p_2\, f(n + 1, x + \eta - \delta/2) + (1 - p_2)\, f(n + 1, x - \eta + \delta/2)}{1 + r_*},$$

$$f(n, x - \delta/2) = \frac{q_2\, f(n + 1, x + \eta - \delta/2) + (1 - q_2)\, f(n + 1, x - \eta + \delta/2)}{1 + r_*},$$

and here the answer is given as

$$\mathbb{C}^* - f(0, x_0 - \delta/2)$$

$$\gamma_0 = \frac{f(1, x_0 + \eta - \delta/2) - (1 - r_*)\, f(0, x_0 - \delta/2)}{(\eta + \alpha + \theta)(1 + r_*)},$$

$$\beta_0 = -f(0, x_0 - \delta/2) - \gamma_0\,(x_0 - \delta/2).$$

Analogous computations can be carried out (with obvious changes in the formulas) for a put option.

Below we present graphs of computer calculations of the ask and bid prices $\mathbb{C}^*$ and $\mathbb{C}_*$, along with the hedging strategies obtained on the basis of the "backward induction" algorithm.[1] The horizontal axis shows the time (5 discrete steps), and the vertical axis shows the stock price. The nodes of the binomial lattice are the times for making decisions about changing the hedging portfolio. The hedging strategy is represented in the format *capital* (rubles) (*number of stocks, number of bonds*), where a negative number of stocks indicates leasing of stocks, and a negative number of bonds indicates borrowing.

---

[1]Developed by K. M. Feoktistov.

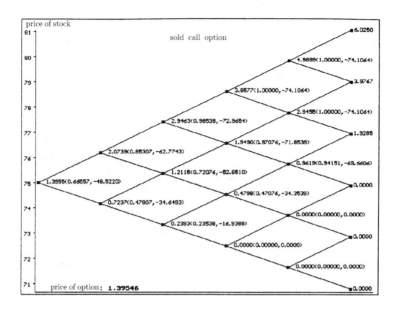

FIG. 1. Pricing a call option for an "ideal" model

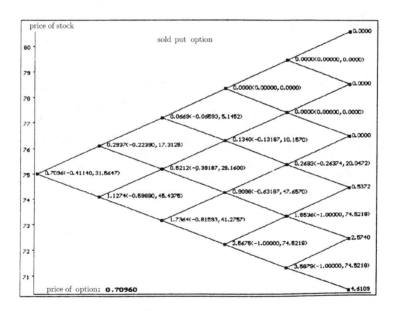

FIG. 2. Pricing a put option for an "ideal" model

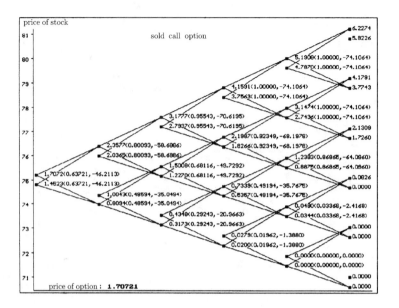

FIG. 3. Pricing a call option with spread taken into account

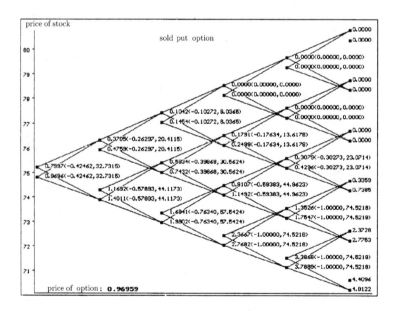

FIG. 4. Pricing a put option with spread taken into account

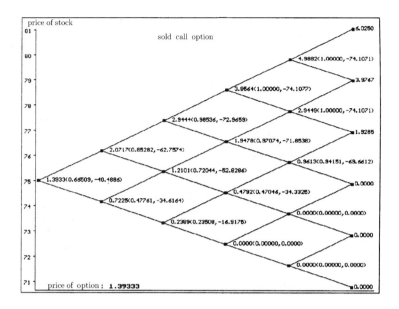

FIG. 5. Pricing a call option with dividends taken into account

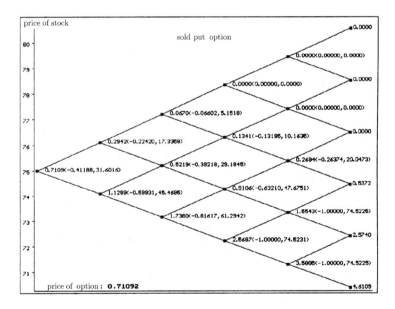

FIG. 6. Pricing a put option with dividends taken into account

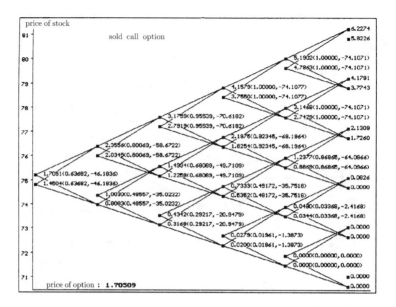

FIG. 7. Pricing a call option with spread and leasing taken into account

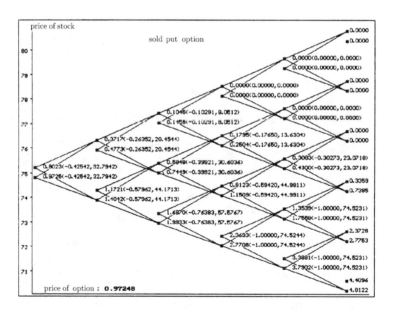

FIG. 8. Pricing a put option with spread and leasing taken into account

Appendix 2

This appendix[1] illustrates the use of the methods of Lecture 8 for solving investment problems (see § 8.2).

We consider the *binomial* model of a (B, S)-market. Let the sequences $B = (B_n)_{n \geq 0}$ and $S = (S_n)_{n \geq 0}$ give the prices of bonds and the prices of some asset, respectively. Suppose that

$$\Delta B_n = r\, B_{n-1}, \qquad r = \text{const}, \quad n = 1, \ldots, N,$$
$$B_0 = 1,$$
$$\Delta S_n = \rho_n\, S_{n-1}, \qquad n = 1, \ldots, N,$$
$$S_0 = 5,000,$$

where $\{\rho_n\}_{n \geq 1}$ is a sequence of independent identically distributed random variables, and

$$\mathbf{P}\{\rho_n = b\} = p,$$
$$\mathbf{P}\{\rho_n = a\} = 1 - p.$$

The evolution of the investor's capital is given by the formulas (3.5).

We let $r = 0.0007$, which will correspond to a bond yield of approximately 30% per year, $a = -0.0003$, $b = 0.0015$, and $p = 0.5$. An optimal strategy $\widehat{\pi} = (\widehat{\pi}_n)_{n \geq 1}$ for an investor with initial capital $x = 100,000$ and investment goal $f = 110,000$ is given by the formulas (8.22) and (8.23). They can be rewritten in the framework of this model as

$$\widehat{\gamma}_n = \frac{\mu\, B_n}{\sigma^2 S_{n-1}} \left(\frac{f}{B_N} - \frac{X_{n-1}^{\widehat{\pi}}}{B_{n-1}} \right),$$
$$\widehat{\beta}_n = \frac{X_{n-1}^{\widehat{\pi}}}{B_{n-1}} - \frac{\mu\,(1+r)}{\sigma^2} \left(\frac{f}{B_N} - \frac{X_{n-1}^{\widehat{\pi}}}{B_{n-1}} \right),$$

where

$$\mu = \mathbf{E}\,(\rho_n - r) = 2 \cdot 10^{-4},$$
$$\sigma^2 = \mathbf{E}\,(\rho_n - r)^2 = 4 \cdot 10^{-7}.$$

Below we give graphs of the evolution of the capital controlled by the strategy $\widehat{\pi} = (\widehat{\pi}_n)_{n \geq 1}$, for a portfolio consisting of bonds, with

$$\beta_n \equiv \frac{X_0}{B_0} = \text{const},$$
$$\gamma \equiv 0, \qquad n = 1, \ldots, N,$$

---

[1]Prepared by M. L. Nechaev.

and for a portfolio consisting of stocks, with

$$\beta_n \equiv 0,$$
$$\gamma \equiv \frac{X_0}{B_0} = \text{const}, \qquad n = 1, \ldots, N.$$

The graphs are modeled according to the model parameters chosen above. Figures 9–11 indicate also the boundaries inside which the value corresponding to the strategy $\widehat{\pi} = (\widehat{\pi}_n)_{n \geq 1}$ must stay with probability 0.9.

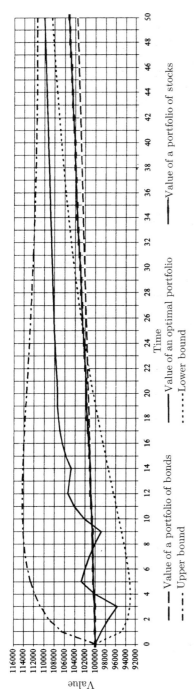

FIG. 9 Graph of the evolution of the value of an optimal portfolio
(one of the possible realizations, constructed by computer simulation)

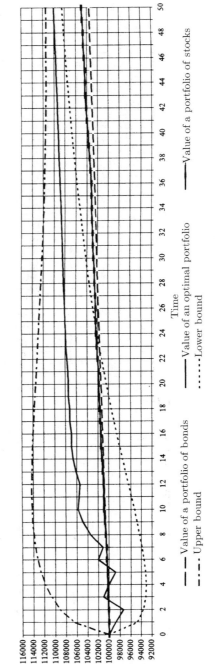

FIG.10 Graph of the evolution of the value of an optimal portfolio
(one of the possible realizations, constructed by computer simulation)

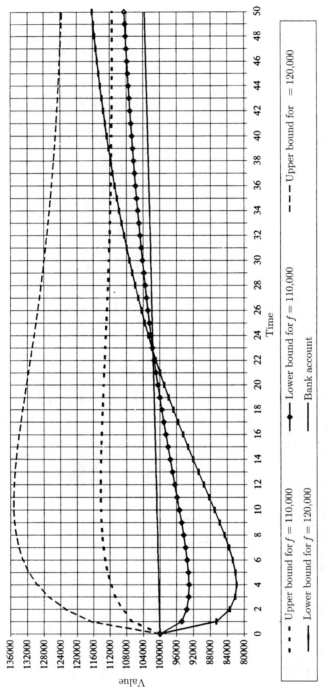

FIG. 11 Lower and upper bounds, within which the value of the optimal portfolio stays with probability greater than 0.9, for different values of the investment goal f

Appendix 3[*]

QUESTIONS OF NO-ARBITRAGE AND COMPLETENESS OF
DISCRETE MARKETS AND PRICING OF CONTINGENT CLAIMS

We consider a general model of a discrete market by a finite number of risky
assets taking finitely many values. For this model we investigate criteria for no-
arbitrage and completeness of the market, and we study a method for pricing contingent claims. Special attention is given to characterizing prices in terms of completions of the original market.

Multinomial markets. The absence of arbitrage and completeness

Let us consider a market with a single nonrisky asset called a *bond* and $L \geq 1$
risky assets called *stocks*. The behavior of the market is described by an $(L+1)$-
dimensional stochastic sequence $S = (S_t : t = 0, \ldots, T)$, where the lth component
S_t^l of $S_t = (S_t^0, \ldots, S_t^L)^{\mathrm{T}}$ is the price of the asset l. A finite probability space
$(\Omega, \mathcal{F}, \mathbf{P})$ is given ($|\Omega| < \infty$) with probability measure $\mathbf{P}$ such that $\mathbf{P}(\{\omega\}) > 0$ for
all $\omega \in \Omega$.

The prices generate a natural filtration $(\mathcal{F}_t : t = 0, \ldots, T)$, where $\mathcal{F}_t =
\sigma(S_0, \ldots, S_t)$, $\mathcal{F}_0 = \{\varnothing, \Omega\}$, and $\mathcal{F}_T = 2^\Omega$. Corresponding to each $\mathcal{F}_t$ is a min-
imal measurable partition $\mathbb{P}_t$ of the space Ω. Thus, at the time t the investor knows
only the element of $\mathbb{P}_t$ containing ω.

It can be assumed without loss of generality that $S_t^0 \equiv 1$ for all t, since otherwise
we could pass to the sequence of discounted prices $\overline{S}_t^l = S_t^l/S_t^0$.

A strategy (portfolio) is defined to be a predictable $(L+1)$-dimensional sto-
chastic sequence $\pi = (\pi_t : t = 0, \ldots, T)$, where $\pi_t = (\gamma_t^0, \gamma_t^1, \ldots, \gamma_t^L)$, and γ_t^l is the
amount of the asset l in the portfolio in the period $[t-1, t)$. The portfolio changes
at the times $t = 0, \ldots, T-1$. The value X_t^π of the strategy π at the time t is equal
to the inner product $\pi_t S_t$.

A portfolio is said to be *self-financing* if $\pi_t S_t = \pi_{t+1} S_t$ for all $t = 1, \ldots, T-1$.
A market is called a *no-arbitrage* market if there is no self-financing strategy π such
that:

(a) $X_0^\pi = 0$;
(b) $X_t^\pi \geq 0$ for all $t = 0, \ldots, T$ and $\omega \in \Omega$;
(c) $X_T^\pi \not\equiv 0$ on Ω.

In investigating many of the properties of a market it often suffices to consider
a single step of the process, and the general result is obtained by the scheme of
backward induction. At time t the investor knows only the element A of $\mathbb{P}_t$ containing the process. Let $A_1, \ldots, A_K$ be the elements of the partition $\mathbb{P}_{t+1}$ lying

---

[*]Prepared jointly with K. M. Feoktistov.

in A. Thus, in the step $t+1$ the process passes from the state A to one of the states $A_1, \ldots, A_K$ with positive probabilities $\mathbf{P}(A_i \mid A)$. Letting $\mathcal{A}_A = \sigma(A_1, \ldots, A_K)$ and $\mathbf{P}_A(\cdot) = \mathbf{P}(\cdot \mid A)$ for the elements A, we get the reduced one-step probability space $(A, \mathcal{A}_A, \mathbf{P}_A)$.

The strategy at the step t depends only on the current situation on the market (A) and on the $\mathcal{A}_A$-measurable behavior of the process in the single step $\Delta S_{t+1} = (S_{t+1}^0 - S_t^0, \ldots, S_{t+1}^L - S_t^L)^{\mathrm{T}}$. We use the following notation:

$$\mathrm{conv}(M) = \left\{ x : x = \sum_{x_m \in M} \lambda_m x_m, \ \lambda_m \geq 0, \ \sum \lambda_m = 1 \right\},$$

the convex hull of a set M;

$$\mathrm{aff}(M) = \left\{ x : x = \sum_{x_m \in M} \lambda_m x_m, \ \sum \lambda_m = 1 \right\},$$

the affine span of M;

$$\mathrm{ri}(M) = \{ x : B_\varepsilon(x) \cap \mathrm{aff}(M) \subset M \text{ for some } \varepsilon > 0 \},$$

the *relative interior* of M as a subset of $\mathrm{aff}(M)$, where $B_\varepsilon(x)$ is the ball about x of radius ε;

$$S_t(A) = \{ S_t(\omega) : \omega \in A \} = (S_t^0(A), \ldots, S_t^L(A))^{\mathrm{T}},$$

the price vector of the assets at time t on an element $A \in \mathbb{P}_t$;

$$\{ S_{t+1}(A) \} = \{ S_{t+1}(\omega) : \omega \in A \} = \{ S_{t+1}(A_1), \ldots, S_{t+1}(A_m) \},$$

the set of price vectors of the assets at time $t+1$ on the elements $A_k \in \mathbb{P}_{t+1}$, $k = 1, \ldots, K$, or $k = 1, \ldots, m$ if the measure $\mathbf{Q}_A \not\sim \mathbf{P}_A$ and gives a positive probability only on $A_1, \ldots, A_m$;

$$S_{t+1}(A) = (S_{t+1}(A_1), \ldots, S_{t+1}(A_m)),$$

the price matrix of the assets at the time $t+1$, which has the matrix expression

$$\mathbb{S} = S_{t+1}(A) = \begin{pmatrix} S_{t+1}^0(A_1) & \cdots & S_{t+1}^0(A_m) \\ \vdots & \ddots & \vdots \\ S_{t+1}^L(A_1) & \cdots & S_{t+1}^L(A_m) \end{pmatrix};$$

$$\mathbf{E}_{\mathbf{Q}_A}(\Delta S_{t+1}(A)) = (\Delta S_{t+1}(A_1), \ldots, \Delta S_{t+1}(A_m))(q_1, \ldots, q_m)^{\mathrm{T}}$$
$$= \mathbf{E}_{\mathbf{Q}_A}(\Delta S_{t+1} \cdot \mathbf{1}_A).$$

THEOREM 1 (no-arbitrage criterion). *The following conditions are equivalent*:
(a) *the market is a no-arbitrage market*;
(b) $\mathbf{0} \in \mathrm{ri}\,\mathrm{conv}\{\Delta S_{t+1}(A)\}$ *for all* $t = 0, \ldots, T$ *and all* $A \in \mathbb{P}_t$;
(c) *there is a probability measure* $\mathbf{Q} \sim \mathbf{P}$ *such that* S *is a* $(\mathbf{Q}, \mathcal{F}_t)$*-martingale*.

Along with this criterion we need the following matrix formulation of it, which is immediately equivalent to (c) for a pair (t, A): *a market is a no-arbitrage market if and only if the system*

$$(1) \qquad \begin{cases} (S_{t+1}(A))\mathbf{p} = S_t(A), \\ p_K > 0 \quad \forall k = 1, \ldots, K \end{cases}$$

has a solution for all $t = 0, \ldots, T$ *and* $A \in \mathbb{P}_t$.

A *contingent claim* f_T is defined to be any $\mathcal{F}_t$-measurable vector-valued function. A market is said to be *complete* if for any contingent claim f_T there is a self-financing strategy π such that $X_T^\pi = f_T$ **P**-a.s. ($\Leftrightarrow \forall \omega \in \Omega$).

A market is said to be *complete with respect to a measure* **Q** that is not necessarily equivalent to **P** if for any contingent claim f_T there is a strategy π such that $X_T^\pi = f_T$ **Q**-a.s. ($\Leftrightarrow \forall \omega \in \Omega$ with $\mathbf{Q}(\{\omega\}) > 0$).

A probability measure **Q** will be called a *martingale measure* if the stochastic sequence S is a martingale with respect to it. Let $\mathbb{P}^*$ denote the set of martingale measures equivalent to **P**, let $\mathbb{M}(S)$ denote the set of martingale measures (not necessarily equivalent to **P**), and let $\mathbb{M}_e(S)$ denote the set of extremal points of $\mathbb{M}(S)$.

ASSERTION 1 (see [**40**], Theorem 4.1). *Let* $\mathbf{Q} \in \mathbb{M}(S)$. *Then* $\mathbf{Q} \in \mathbb{M}_e(S) \Leftrightarrow$ *the market is complete with respect to* **Q** (*that is, the measures with respect to which the market is complete are the extremal points of the set of all martingale measures*).

The completeness criterion has the following form for a one-step reduced model.

ASSERTION 2. *The following conditions are equivalent*:

(a) $\mathbf{Q}_A$ *is an extremal point of the set* $\{\mathbf{Q}_A' : \mathbf{E}_{\mathbf{Q}_A'}(\Delta S_{t+1}(A)) = 0\}$;

(b) $\dim\{\mathrm{span}\{\Delta S_{t+1}(A)\}\} = \dim\{\mathrm{span}\{\Delta S_{t+1}(A_1), \ldots, \Delta S_{t+1}(A_m)\}\} = m-1$, *where the measure* $\mathbf{Q}_A$ *gives a positive probability to the elements* $A_1, \ldots, A_m$ *and is zero on the other elements*;

(c) *any claim given on* $A_1, \ldots, A_m$ *is strictly attainable*.

Since $S_t^0(A) \equiv 1$, the quantity $S_t(A)$ is linearly independent of $\{\Delta S_{t+1}(A_1), \ldots, \Delta S_{t+1}(A_m)\}$, and

$$\dim\{\mathrm{span}\{S_{t+1}(A_1), \ldots, S_{t+1}(A_m)\}\}$$
$$= \dim\{\mathrm{span}\{S_{t+1}(A), \Delta S_{t+1}(A_1), \ldots, \Delta S_{t+1}(A_m)\}\}$$
$$= \dim\{\mathrm{span}\{\Delta S_{t+1}(A_1), \ldots, \Delta S_{t+1}(A_m)\}\} + 1.$$

It is clear that on $\mathcal{A}_A$ the condition (b) is equivalent to the relation:

$$\dim\{\mathrm{span}\{S_{t+1}(A)\}\} = \dim\{\mathrm{span}\{S_{t+1}^0, \ldots, S_{t+1}^L\}\} = m,$$

and then for any claim $f = (f_1, \ldots, f_m) \in \mathbf{R}^m$ the system of m equations

$$\boldsymbol{\pi}_t \mathbb{S}_{t+1}(A) = f$$

has a solution and determines a strategy replicating the claim f.

Following [**29**], we define the *splitting index* to be the quantity

$$K(t, A) = |\{A' : A' \in \mathbb{P}_{t+1}, \ A' \subset A\}|,$$

which is very useful in characterizing the properties of completeness of a market.

THEOREM 2 (completeness criterion; see [**40**], Theorem 4.2). *A no-arbitrage market is complete if and only if* $K(t, A) = \dim\{\mathrm{span}\{S_{t+1}(A)\}\}$ *for all* $t = 0, \ldots, T - 1$ *and* $A \in \mathbb{P}_t$. *If here completeness with respect to a nonequivalent measure* $\mathbf{Q}_A$ *is being considered, then the splitting index is defined to be*

$$K(t, A) = |\{A' : A' \in \mathbb{P}_{t+1}, \ A' \subset A, \ \mathbf{Q}_A(A') > 0\}|.$$

Since $\mathrm{rk}\, S_{t+1}(A) = \dim\{\mathrm{span}\{S_{t+1}(A)\}\}$, Theorem 2 admits a convenient matrix formulation:

A no-arbitrage market is complete if and only if

$$\operatorname{rk} S_{t+1}(A) = K(t,A) \quad \text{for all } t = 0,\dots,T \text{ and } A \in \mathbb{P}_t.$$

We remark that under this condition the system (1) has a unique solution. But if the matrix $S_{t+1}(A)$ is square (and thus $K(t,A) = L+1$), then this condition becomes the inequality $\dim S_{t+1}(A) \neq 0$.

Markets with independent assets

What form do no-arbitrage and completeness criteria take for a financial market with independent assets? As before, one of the assets is nonrisky (a *bond*) and $L \geq 1$ of them are risky (*stocks*). We rewrite the model for evolution of the prices $(S_t^0, \dots, S_t^L)$ of the assets as a system of $L+1$ discrete stochastic equations:

$$\begin{cases} \Delta S_t^0 = \rho_t^0 S_{t-1}^0, \\ \vdots \\ \Delta S_t^L = \rho_t^L S_{t-1}^L, \end{cases}$$
$$t = 0, \dots, T, \quad S_0^0, \dots, S_0^L = 1.$$

The zero asset S^0 — the bond — is a predictable process. As earlier, we assume that $S_t^0 \equiv 1$ for all t, and hence $\rho_t^0 \equiv 0$. The remaining random variables ρ_t^l are assumed to be jointly independent. Assuming (to simplify the notation) that the behavior of the process does not change with time, we denote the set of outcomes of the growth rate ρ_t^l of the lth asset by $\boldsymbol{\rho}^l = \{a_1^l, \dots, a_{R^l}^l\}$, where $R^l \geq 2$ and

$$-1 \leq a^l = a_1^l \leq \cdots \leq a_{R^l}^l = b^l, \quad l = 1, \dots, L.$$

It is clear that for this model, Ω consists of finitely many elements:

$$|\Omega| = \left(\prod_{l=1}^{L} R^l \right)^N.$$

The no-arbitrage criterion reduces to the conditions

$$\mathbf{0} \in \operatorname{ri} \operatorname{conv} \{ \Delta S_{t+1}(A) \} \quad \text{for all } t = 0, \dots, T \text{ and } A \in \mathbb{P}_t.$$

Further, for fixed t and A

$$\mathbf{0} \in \operatorname{ri} \operatorname{conv} \{ \Delta S_{t+1}(A) \} = \operatorname{ri} \operatorname{conv} \{ (S_t^0(A)\rho_{t+1}^0(A), \dots, S_t^L(A)\rho_{t+1}^L(A))^{\mathrm{T}} \}$$
$$\Leftrightarrow \mathbf{0} \in \operatorname{ri} \operatorname{conv} \{ (\rho_{t+1}^0(A), \dots, \rho_{t+1}^L(A))^{\mathrm{T}} \}.$$

Since $\{\rho_{t+1}^l(A)\} = \boldsymbol{\rho}^l$, it follows that $\operatorname{conv} \boldsymbol{\rho}^l = [a^l, b^l]$, and

$$\mathbf{0} \in \operatorname{ri} \{ \{0\} \times [a^1, b^1] \times \cdots \times [a^L, b^L] \} = \{0\} \times (a^1, b^1) \times \cdots \times (a^L, b^L)$$
$$\Leftrightarrow a^l < 0 < b^l, \quad l = 1, \dots, L.$$

In the case of a nonzero nonrisky interest rate we obtain an analogous expression for the discounted quantities by passing to the discounted process $\overline{S}_t^l = S_t^l / S_t^0$:

$$\overline{a}^l < 0 < \overline{b}^l, \quad l = 1, \dots, L.$$

Returning to the original prices, we get that

$$\overline{S}_t^l = \frac{S_t^l}{S_t^0} \implies (1+\overline{\rho}_{t+1}^l)\overline{S}_{t+1}^l = \frac{(1+\rho_{t+1}^l)S_t^l}{(1+r)S_t^0} \implies \begin{cases} \overline{a}^l = \frac{1+a^l}{1+r} - 1, \\ \overline{b}^l = \frac{1+b^l}{1+r} - 1 \end{cases}$$

for $l = 1, \ldots, L$, and

$$\frac{1+a^l}{1+r} - 1 < 0 < \frac{1+b^l}{1+r} - 1 \quad \forall l = 1, \ldots, L \Leftrightarrow a^l < r < b^l \quad \forall l = 1, \ldots, L.$$

The result of the above is the following assertion.

ASSERTION 3. *Such a market with independent assets is a no-arbitrage market if and only if $a^l < r < b^l$ for all $l = 1, \ldots, L$ for any $t = 1, \ldots, T$ and $A \in \mathbb{P}_t$.*

To study the completeness of this market we consider the general condition for completeness

$$K(t, A) = \dim \operatorname{span}\{S_{t+1}(A)\},$$

and thus

$$K(t, A) = \prod_{l=1}^{L} R_l \geq 2^L.$$

Since $S_{t+1}(\omega) \in \mathbf{R}^{L+1}$, it follows that

$$\dim \operatorname{span}\{S_{t+1}(A)\} \leq L + 1,$$

and this leads to the necessary condition $L + 1 \geq 2^L$ for completeness, that is, $L = 1$. This at once implies another condition:

$$2 = L + 1 \geq \dim \operatorname{span}\{S_{t+1}(A)\} = K(n, A) = R_1 \geq 2^1, \quad \text{and } R_1 = 2.$$

THEOREM 3 (completeness criterion for a market with independent assets). *A discrete no-arbitrage market with independent assets is complete if and only if it consists solely of one nonrisky asset and one binomial risky asset, that is, the unique model of a complete discrete market with independent assets is the Cox–Ross–Rubinstein model.*

Pricing on incomplete markets

We return to consideration of a general discrete market, on which we must compute the initial price of a given contingent claim $f_T = f_T(S_T) = f_T(S_T^1, \ldots, S_T^L)$. This problem is first solved on a reduced probability space $(A, \mathcal{A}_A, \mathbf{P}_A)$, and in the general case the method of backward induction is applicable.

Let $t = T - 1$ and $A \in \mathbb{P}_t$. The cost of the claim is reduced to $(A, \mathcal{A}_A, \mathbf{P}_A)$ as follows:

$$f_{t+1}(A_k) = f_T(S_T(A_k)), \qquad k = 1, \ldots, K, \quad t = T - 1,$$

or, in vector form,

$$f_{t+1} = f_{t+1}(A) = (f_{t+1}(A_1), \ldots, f_{t+1}(A_K)) = (f_T(A_1), \ldots, f_T(A_K)).$$

The exercise price $\mathbb{C}(t, A)$ $(A \in \mathbb{P}_t)$ obtained in solving this one-step problem is taken as the price of the claim at the next step for smaller t:

$$f_{t+1}(A_k) = \mathbb{C}(t+1, A_k), \qquad A_k \in \mathbb{P}_{t+1}, \quad k = 1, \ldots, K, \quad t = T - 1,$$

$$f_{t+1} = f_{t+1}(A) = (f_{t+1}(A_1), \ldots, f_{t+1}(A_K)) = (\mathbb{C}(t+1, A_1), \ldots, \mathbb{C}(t+1, A_K)),$$

$$k = 1, \ldots, K, \quad t = T - 2, \ldots, 0.$$

The price $\mathbb{C}(0, \Omega)$ obtained at the last step $t = 0$ will be the desired price $\mathbb{C}$.

The market is incomplete, so there is not a unique fair price. On this market there is a closed interval of no-arbitrage prices between the bid price $\mathbb{C}_*$ and the ask price $\mathbb{C}^*$, which are defined by

$$\mathbb{C}_* = \sup_{\pi \in \Pi_*} X_0^\pi, \qquad \text{where} \quad \Pi_* = \{\pi : \pi \in \mathrm{SF}, \ X_T^\pi \leq f_T \ \mathbf{P}\text{-a.s.}\}$$

$$\mathbb{C}^* = \inf_{\pi \in \Pi^*} X_0^\pi, \qquad \text{where} \quad \Pi^* = \{\pi : \pi \in \mathrm{SF}, \ X_T^\pi \geq f_T \ \mathbf{P}\text{-a.s.}\}.$$

The bid and ask prices are defined similarly on a reduced space: namely, letting $\pi = \pi_{t+1} = (\gamma_{t+1}^0, \ldots, \gamma_{t+1}^L)$ be a portfolio bought at the time t at the price $S_t(A)$ and sold at the time $t + 1$ at the price $S_{t+1}(A)$, we get that

$$\mathbb{C}_*(t, A) = \sup_{\pi \in \Pi_*(t, A)} X_t^\pi, \qquad \text{where} \quad \Pi_*(t, A) = \{\pi : X_{t+1}^\pi(A) \leq f_{t+1}(A)\}$$

$$\mathbb{C}^*(t, A) = \inf_{\pi \in \Pi^*(t, A)} X_t^\pi, \qquad \text{where} \quad \Pi^*(t, A) = \{\pi : X_{t+1}^\pi(A) \geq f_{t+1}(A)\}.$$

Substituting

$$X_t^\pi = \boldsymbol{\pi}_t S_t(A),$$

$$\mathbb{S}_{t+1} = S_{t+1}(A)$$

in the above formulas, we get that

$$\begin{cases} \mathbb{C}^* = \boldsymbol{\pi}_t S_t(A) \to \min, \\ \boldsymbol{\pi}_t \mathbb{S}_{t+1} \geq f_{t+1}(A), \end{cases}$$
$$\begin{cases} \mathbb{C}_* = \boldsymbol{\pi}_t S_t(A) \to \max \\ \boldsymbol{\pi}_t \mathbb{S}_{t+1} \leq f_{t+1}(A). \end{cases}$$

Thus, the ask and bid prices are solutions of linear programming problems. Constructing the dual problems, we come to another way of finding the ask and bid prices:

(2)
$$\begin{cases} \mathbb{C}^* = f_{t+1}(A)\mathbf{q} \to \max, \\ \mathbb{S}_{t+1}\mathbf{q} = S_t(A), \\ q_k \geq 0, \quad k = 1, \ldots, K, \end{cases}$$
$$\begin{cases} \mathbb{C}_* = f_{t+1}(A)\mathbf{q} \to \min, \\ \mathbb{S}_{t+1}\mathbf{q} = S_t(A), \\ q_k \geq 0, \quad k = 1, \ldots, K, \end{cases}$$

where $\mathbf{q} = (q_1, \ldots, q_k)^\mathrm{T}$.

These problems are problems of finding the maximum and minimum of the mathematical expectation of the payment function over all martingale measures:

(3)
$$\mathbb{C}^* = \sup_{q \in \mathbb{M}(S)} \mathbf{E}_q(f_{t+1}(A)),$$
$$\mathbb{C}_* = \inf_{q \in \mathbb{M}(S)} \mathbf{E}_q(f_{t+1}(A)).$$

The above system of equations with the constraints

$$\mathbb{S}_{t+1}\mathbf{q} = \mathbf{E}_q(S_{t+1}(A))$$
$$= \begin{pmatrix} S_{t+1}^0(A_1) & \cdots & S_{t+1}^0(A_K) \\ \vdots & \ddots & \vdots \\ S_{t+1}^L(A_1) & \cdots & S_{t+1}^L(A_K) \end{pmatrix} \begin{pmatrix} q_1 \\ \vdots \\ q_K \end{pmatrix} = \begin{pmatrix} S_t^0(A) \\ \vdots \\ S_t^K(A) \end{pmatrix} = S_t(A)$$

gives a condition for the martingale property with respect to the measure $\mathbf{q}$. The system of inequalities together with the first equation of the system, which has the form $q_1 + \cdots + q_K = 1$, shows that $\mathbf{q}$ is a probability measure defined on the reduced space $(A, \mathcal{A}_A, \mathbf{P}_A)$.

ASSERTION 4. *The infimum and supremum in* (3) *can be taken over the set* $\mathbb{P}^*$ *of martingale measures equivalent to* $\mathbf{P}$ *instead of over the set* $\mathbb{M}(S)$ *of all martingale measures. Further, if the infimum is different from the supremum, then neither is attained in the class of measures equivalent to* $\mathbf{P}$.

PROOF. 1. The sets $\mathbb{M}(S)$ and $\mathbb{P}^*$ are convex, since they are specified by linear equations and inequalities.

2. We prove that $\mathbb{M}(S) = \overline{\mathbb{P}^*}$: if $\mathbf{q} \in \mathbb{M}(S) \setminus \overline{\mathbb{P}^*}$ and $\mathbf{p} \in \mathbb{P}^* \subseteq \mathbb{M}(S)$, then

$$\lambda\mathbf{q} + (1 - \lambda)\mathbf{p} \in \mathbb{M}(S) \quad \forall \lambda \in [0, 1],$$
$$\lambda\mathbf{q} + (1 - \lambda)\mathbf{p} \in \mathbb{P}^* \quad \forall \lambda \in [0, 1),$$

and hence $\mathbf{q}$ is a limit point of the set $\mathbb{P}^*$, so that $\mathbf{q} \in \overline{\mathbb{P}^*}$.

3. The supremum (infimum) over a set is equal to the supremum (infimum) over the closure of the set.

4. Since (2) is a linear programming problem if the supremum is not equal to the infimum, the region of feasible solutions is a simplex of dimension greater than zero, and the solution is attained on its boundary, where one of the q_i is equal to zero. Consequently, the solution is not attained on the set of equivalent measures.

Thus,

$$\mathbb{C}_*(t, A) = \sup_{\pi \in \Pi_*(t,A)} X_t^\pi = \inf_{\mathbf{q} \in \mathbb{P}^*} \mathbf{E}_q(f_{t+1}(A))$$
$$\leq \sup_{\mathbf{q} \in \mathbb{P}^*} \mathbf{E}_{\mathbf{q}}(f_{t+1}(A)) = \inf_{\pi \in \Pi_*(t,A)} X_t^\pi = \mathbb{C}^*(t, A)$$

on the reduced probability space for any payment functions. The same relations are valid also for the original probability space, as shown by the next assertion.

ASSERTION 5. *Finding the supremum* (infimum) *of the mathematical expectation of the payment function over all probability measures on the original space is equivalent to constructing a probability measure from the reduced spaces* (by backward induction), *with the supremum* (infimum) *found for each reduction separately.*

An analogous assertion is true also for strategies.

PROOF. 1. The supremum of the expectation over all measures is greater than or equal to the expectation with respect to any measure, in particular, the one constructed with respect to the reduced measures.

2. Suppose that $\mathbf{Q} = \arg \sup_{\mathbf{Q} \in \mathbb{P}^*} \mathbf{E}_{\mathbf{Q}} f_T$ but that there is a reduced space with "index" (t, A) such that $\mathbf{E}_{\mathbf{q}}(f_{t+1}(A))$ does not attain its maximum, as a function of $\mathbf{q}$, on the measure $q_k(t, A) = \mathbf{Q}(A_k \mid A)$. We define a new reduced measure $\mathbf{q}'(t, A) = \arg \sup_{\mathbf{q} \in \mathbb{P}^*} \mathbf{E}_{\mathbf{q}}(f_{t+1}(A))$ and construct a measure on the original space with respect to the set of reduced measures, with only $\mathbf{q}$ replaced by $\mathbf{q}'$:

$$\mathbf{Q}(\omega) = \mathbf{Q}(A_{k_1 k_2 \ldots k_T}) \mathbf{Q}(A_{k_T} \mid A_{k_1 \ldots k_{T-1}}) \mathbf{Q}(A_{k_1 \ldots k_{T-1}}) = \cdots$$
$$= \mathbf{Q}(A_{k_T} \mid A_{k_1 \ldots k_{T-1}}) \mathbf{Q}(A_{k_{T-1}} \mid A_{k_1 \ldots k_{T-2}}) \cdots \mathbf{Q}(A_{k_2} \mid A_{k_1}) \mathbf{Q}(A_{k_1} \mid \Omega)$$
$$= q_{k_T}(T - 1, A_{k_1 \ldots k_{T-1}}) q_{k_{T-1}}(T - 2, A_{k_1 \ldots k_{T-2}}) \cdots q_{k_1}(0, \Omega).$$

As a result of the substitution, we get that

$$\mathbf{E}_{\mathbf{Q}}(f_{\tau+1}(B) \mid \mathcal{F}_\tau) = \mathbf{E}_{\mathbf{Q}'}(f_{\tau+1}(B) \mid \mathcal{F}_\tau) \quad \forall \tau > t, \ \forall B \in \mathbb{P}_\tau,$$
$$\mathbf{E}_{\mathbf{Q}}(f_{t+1}(A) \mid \mathcal{F}_t) < \mathbf{E}_{\mathbf{Q}'}(f_{t+1}(A) \mid \mathcal{F}_t),$$
$$\mathbf{E}_{\mathbf{Q}}(f_T \mid \mathcal{F}_t) \leq \mathbf{E}_{\mathbf{Q}'}(f_T \mid \mathcal{F}_t),$$
$$\mathbf{E}_{\mathbf{Q}}(f_T) = \mathbf{E}_{\mathbf{Q}}(\mathbf{E}_{\mathbf{Q}}(f_T \mid \mathcal{F}_\tau)) \leq \mathbf{E}_{\mathbf{Q}'}(\mathbf{E}_{\mathbf{Q}'}(f_T \mid \mathcal{F}_\tau)) = \mathbf{E}_{\mathbf{Q}'}(f_T),$$

so that $\mathbf{Q}' = \arg \sup_{\mathbf{Q} \in \mathbb{P}^*} \mathbf{E}_{\mathbf{Q}} f_T$.

The remaining assertions are proved according to the same scheme.

Assertions 4 and 5 immediately yield the following.

THEOREM 4. *The ask and bid prices of any contingent claim* $f_T = f_T(S_T) = f_T(S_T^1, \ldots, S_T^L)$ *satisfy*

$$\mathbb{C}_* = \sup_{\pi \in \Pi_*} X_0^\pi = \inf_{\mathbf{Q} \in \mathbb{P}^*} \mathbf{E}_{\mathbf{Q}}(f_T) \leq \sup_{\mathbf{Q} \in \mathbb{P}^*} \mathbf{E}_{\mathbf{Q}}(f_T) = \inf_{\pi \in \Pi^*} X_0^\pi = \mathbb{C}^*,$$

where

$$\Pi^* = \{\pi : \pi \in \mathrm{SF}, \ X_T^\pi \geq f_T \ \mathbf{P}\text{-}a.s.\},$$
$$\Pi_* = \{\pi : \pi \in \mathrm{SF}, \ X_T^\pi \leq f_T \ \mathbf{P}\text{-}a.s.\},$$

and $\mathbb{P}^*$ *is the set of martingale measures equivalent to* $\mathbf{P}$.

This theorem is an improvement of the result obtained in Lecture 7, under the assumption that the function f_T is convex.

Completion of markets

We now raise the question of whether it is possible to complete the original incomplete market. Since the completeness criterion has the form

$$(4) \qquad\qquad K(t, A) = \dim \operatorname{span} \{S_{t+1}(A)\}$$

in the general case (for a reduced model), it follows from the definition of $K(t, A)$ that

$$K(t, A) = |\{S_{t+1}(A)\}| \geq \dim \operatorname{span} \{S_{t+1}(A)\},$$

and hence it remains to consider the case when

$$K(t, A) > \dim \operatorname{span} \{S_{t+1}(A)\}.$$

Therefore, there are two paths for completion: decrease the left-hand side of (4) or increase the right-hand side.

First path. $K(t, A)$ can be decreased by declaring some outcomes "improbable", that is, as having zero probability. Such a method is presented in [**40**]: transition to a nonequivalent measure $\mathbf{Q}$. Here the right-hand side of (4) remains unchanged. We can find the desired measure $\mathbf{Q} \in \mathbb{M}(S)$ by solving the linear programming problem (2). With the use of Assertion 1 and linear programming theory it is not hard to show that $\mathbf{Q} \in \mathbb{M}_e(S)$ and that on each reduced space the vector $\mathbf{q}$ has exactly $\dim \operatorname{span}\{S_{t+1}(A)\}$ nonzero components.

Second path: increasing the quantity $\dim \operatorname{span}\{S_{t+1}(A)\} \leq L + 1$. Equality holds if and only if the assets are linearly independent as functions on the collection of values $A_1, \ldots, A_K$, since

$$
\dim \operatorname{span}\{S_{t+1}(A)\} = \operatorname{rk} \mathbb{S} = \operatorname{rk} \begin{pmatrix} S_{t+1}^0(A_1) & \cdots & S_{t+1}^0(A_K) \\ \vdots & \ddots & \vdots \\ S_{t+1}^L(A_1) & \cdots & S_{t+1}^L(A_K) \end{pmatrix}.
$$

For example, let

$$
S_{t+1}^L(x) = \sum_{l=0}^{L-1} \mu_l S_{t+1}^l(x) \qquad \text{where} \quad x = A_1, \ldots, A_K.
$$

Then instead of buying a unit of the Lth asset it suffices to buy corresponding portions of the remaining assets. This asset does not play any role on the market, and its removal does not change anything. It will be assumed below that

$$
\dim \operatorname{span}\{S_{t+1}(A)\} = L + 1.
$$

It is clear that for (4) to hold the quantity $\dim \operatorname{span}\{S_{t+1}(A)\}$ must be increased by $K - (L + 1)$. But for this it is necessary to add at least $K - (L + 1)$ linearly independent assets. Here it does not make sense to add an asset that increases the left-hand side of (4) by at least a unit. Consequently, the assets added must be given on the same probability space, that is, they must be functionally dependent on the original assets. We remark that, for example, options of all possible forms or other derivative securities can serve as such functionally dependent but not linearly dependent assets. It is convenient to assume that the splitting index is constant; otherwise it is necessary to vary the number of assets added, and this greatly complicates the notation. We introduce a collection

$$
\begin{cases} S_t^{L+1} = s_t^{L+1}(S_t^0, \ldots, S_t^L), \\ \vdots \\ S_t^{K-1} = s_t^{K-1}(S_t^0, \ldots, S_t^L) \end{cases}
$$

of $K - (L + 1)$ dependent assets such that the no-arbitrage condition holds, and they are linearly independent:

(5)
$$
\det \widetilde{\mathbb{S}} = \det \begin{pmatrix} S_{t+1}^0(A_1) & \cdots & S_{t+1}^0(A_K) \\ \vdots & \ddots & \vdots \\ S_{t+1}^L(A_1) & \cdots & S_{t+1}^L(A_K) \end{pmatrix} \neq 0.
$$

Then (4) holds in the completed market. As mentioned earlier, it does not make sense to add a larger number of assets to complete the market. A proof of the existence of such a minimal completion for an original no-arbitrage market is given in the proof of Assertion 6.

Thus, we get a criterion for a minimal completion of a market.

THEOREM 5 (criterion for a minimal completion of a market). *The collection* $S^{L+1}, \ldots, S^{L'}$ *gives a minimal completion if and only if*

1. $L' = K - 1$;
2. *the assets are dependent, that is,*

$$
\begin{cases}
S_t^{L+1} = s_t^{L+1}(S_t^0, \ldots, S_t^L), \\
\vdots \\
S_t^{L'} = s_t^{L'}(S_t^0, \ldots, S_t^L);
\end{cases}
$$

3. *the completed system*

$$
\begin{cases}
(S_{t+1}(A))\mathbf{p} = S_t(A), \quad t = 0, \ldots, T, \ A \in \mathbb{P}_t, \\
p_k > 0, \quad k = 1, \ldots, K
\end{cases}
$$

of K equations and K inequalities has a unique solution.

Let $\mathbf{M}(S)$ denote the set of minimal completions.

Limit values of the exercise prices upon completion of a market

In completing a market by means of different assets we obtain different costs of a given contingent claim. We determine the bounds of the costs obtained in this way. Notation:

- $\mathbf{M}(S_{t+1}(A)) = \left\{ \widetilde{\mathbb{S}}_{t+1} = \begin{pmatrix} \mathbb{S}_{t+1} \\ \mathbb{X} \end{pmatrix} : \widetilde{\mathbb{S}}_{t+1}\mathbf{q} = \widetilde{S}_t(A) \text{ for a unique } \mathbf{q} > 0 \right\}$ *is the set of completions of the market (on the reduced space), where $\widetilde{\mathbb{S}}_{t+1}$ is a block matrix, and $\mathbb{X}$ is any $(K - L - 1) \times K$ matrix with nonnegative coefficients satisfying the indicated conditions;*
- $\mathbb{P}_{\mathbf{M}}(t+1, A) = \{\mathbf{q} : \widetilde{\mathbb{S}}_{t+1}\mathbf{q} = \widetilde{S}_t(A) \text{ for some } \widetilde{\mathbb{S}}_{t+1} \in \mathbf{M}(S_{t+1}(A)), \ \mathbf{q} > 0\}$ *is the set of martingale measures of completions of the market on the reduced space;*
- $\mathbf{M}(S)$ *is the set of completions of the market;*
- $\mathbb{P}_{\mathbf{M}}\{\mathbf{Q} : \mathbf{Q} \sim \mathbf{P}, \text{ and some } \widetilde{\mathbb{S}} \in \mathbf{M}(S) \text{ is a martingale with respect to } \mathbf{Q}\}$ *is the set of martingale probability measures equivalent to the original measure for all possible completed markets.*

ASSERTION 6.. *On the reduced space the set of martingale measures of all possible completions of the market coincides with the set of equivalent martingale measures of the incomplete market:*

$$
\mathbb{P}_{\mathbf{M}}(t+1, A) = \mathbb{P}_A(t+1, A).
$$

PROOF. 1. $\mathbf{q} \in \mathbb{P}_{\mathbf{M}}(t+1, A) \implies \widetilde{\mathbb{S}}_{t+1}\mathbf{q} = \widetilde{S}_t(A), \mathbf{q} > 0 \implies \mathbb{S}_{t+1}\mathbf{q} = S_t(A),$
$\mathbf{q} > 0 \implies \mathbf{q} \in \mathbb{P}_A(t+1, A).$

2. $\mathbf{q} \in \mathbb{P}_A(t+1, A) \implies \mathbb{S}_{t+1}\mathbf{q} = S_t(A), \mathbf{q} > 0.$

By the theorem on completion of the base, a collection of $L+1$ linearly independent column vectors of the matrix $\mathbb{S}_{t+1}$ can be completed by $K - L - 1$ vectors to form a nonsingular matrix $\widetilde{\mathbb{S}}_{t+1}^i$. Further, the values of $\widetilde{\mathbb{S}}_{t+1}^i$ and $\widetilde{\mathbb{S}}_t^i = \widetilde{\mathbb{S}}_{t+1}^i\mathbf{q}$ can be arbitrary, but by adding a constant to each coordinate of such a vector we can ensure that all coordinates and the expectation are positive with preservation of independence:

$$\widetilde{\mathbb{S}}_{t+1}^i\mathbf{q} = (\widetilde{\mathbb{S}}_{t+1}^i + c\mathbb{S}_{t+1}^0)\mathbf{q} = (\widetilde{\mathbb{S}}_{t+1}^i(A_1) + c, \ldots, \widetilde{\mathbb{S}}_{t+1}^i(A_K) + c)\mathbf{q} > 0,$$

where, for example, $c = -\min_{k=1,\ldots,K}(\widetilde{\mathbb{S}}_{t+1}^i(A_k)) + 1$.

Consequently, $\widetilde{\mathbb{S}}_{t+1}^i(A_k) > 0$ and $\widetilde{\mathbb{S}}_t^i \overset{\text{def}}{=} \widetilde{\mathbb{S}}_{t+1}^i\mathbf{q} > 0$ for all $i = L+1, \ldots, K-1$ and $k = 1, \ldots, K$, which implies the needed relation

$$\widetilde{\mathbb{S}}_{t+1}\mathbf{q} = \widetilde{S}_t(A), \mathbf{q} > 0 \implies \mathbf{q} \in \mathbb{P}_{\mathbf{M}}(t+1, A).$$

THEOREM 6 (fundamental theorem on completion of a market). *The ask and bid prices for any contingent claim* $f_T = f_T(S_T) = f_T(S_T^1, \ldots, S_T^L)$ *satisfy*

$$\mathbb{C}_* = \sup_{\pi \in \Pi_*} X_0^{\pi} = \inf_{Q \in \mathbb{P}^*} \mathbf{E}_Q(f_T) = \inf_{Q \in \mathbb{P}_{\mathbf{M}}} \mathbf{E}_Q(f_T)$$
$$\leq \sup_{Q \in \mathbb{P}_{\mathbf{M}}} \mathbf{E}_Q(f_T) = \sup_{Q \in \mathbb{P}^*} \mathbf{E}_Q(f_T) = \inf_{\pi \in \Pi_*} X_0^{\pi} = \mathbb{C}^*.$$

Further, the exercise costs of the contingent claim that are obtained upon completing the market by different assets lie in the interval $(\mathbb{C}_*, \mathbb{C}^*)$. *(And if* $\mathbb{C}_* = \mathbb{C}^*$, *then all the costs are equal to* $\mathbb{C} = \mathbb{C}_* = \mathbb{C}^*$).

PROOF. 1. Except for the innermost equalities this assertion was proved in Theorem 4.

2. The innermost equalities follow from Assertion 6: on each reduced space the set of martingale measures of the completions of the market coincides with the set of equivalent martingale measures, so the same is true for measures on the original probability space:

$$\mathbb{P} = \mathbb{P}_{\mathbf{M}}.$$

This at once yields the assertion of the theorem.

3. It follows from Assertion 4 that if $\mathbb{C}_* \neq \mathbb{C}^*$, then the infimum and supremum are not attained.

Hints for solving the problems

2.1. Prove the more general fact that $\mathbf{E}(M_{\tau_2} \mid \mathcal{F}_{\tau_1}) = M_{\tau_1}$ for any stopping times $0 \le \tau_1 \le \tau_2 \le N$, and hence $\mathbf{E}M_{\tau_2} = \mathbf{E}M_{\tau_1} = \mathbf{E}M_0$. Consider a set $A \in \mathcal{F}_{\tau_1}$ and the set $B = A \cap \{\omega : \tau_1 = n\}$, and show that

$$\int_A M_{\tau_2}\,d\mathbf{P} = \int_A M_{\tau_1}\,d\mathbf{P}, \quad \text{or} \quad \int_{B \cap \{\tau_2 \ge n\}} M_{\tau_2}\,d\mathbf{P} = \int_{B \cap \{\tau_2 \ge n\}} M_n\,d\mathbf{P}.$$

These equalities are consequences of the following chain of relations, which are valid in view of the martingale property of M:

$$\begin{aligned}
\int_{B \cap \{\tau_2 \ge n\}} M_n\,d\mathbf{P} &= \int_{B \cap \{\tau_2 = n\}} M_n\,d\mathbf{P} + \int_{B \cap \{\tau_2 > n\}} M_n\,d\mathbf{P} \\
&= \int_{B \cap \{\tau_2 = n\}} M_n\,d\mathbf{P} + \int_{B \cap \{\tau_2 > n\}} \mathbf{E}(M_{n+1} \mid \mathcal{F}_n)\,d\mathbf{P} \\
&= \int_{B \cap \{\tau_2 = n\}} M_n\,d\mathbf{P} + \int_{B \cap \{\tau_2 > n\}} M_{n+1}\,d\mathbf{P} \\
&= \int_{B \cap \{n \le \tau_2 \le n+1\}} M_{\tau_2}\,d\mathbf{P} + \int_{B \cap \{\tau_2 \ge n+2\}} M_{n+2}\,d\mathbf{P},
\end{aligned}$$

and so on.

2.2. For any $A \in \mathcal{F}_{N-1}$ the properties of mathematical expectations give us that, on the one hand,

$$\int_A Y\,d\widetilde{\mathbf{P}} = \int_A \widetilde{\mathbf{E}}(Y \mid \mathcal{F}_{N-1})\,d\widetilde{\mathbf{P}} = \int_A \widetilde{\mathbf{E}}(Y \mid \mathcal{F}_{N-1})Z_{N-1}\,d\mathbf{P}$$

and on the other hand,

$$\int_A Y\,d\widetilde{\mathbf{P}} = \int_A Y Z_N\,d\mathbf{P} = \int_A \mathbf{E}(Y Z_N \mid \mathcal{F}_{N-1})\,d\mathbf{P}.$$

2.3. Letting $\Delta\widehat{U}_n = \exp(\Delta U_n) - 1$, show that

$$\exp(U_n) = \exp\left(\sum_{k=1}^n \Delta U_n\right) = \prod_{k=1}^n (1 + \Delta\widehat{U}_k) = \mathcal{E}_N(\widehat{U}).$$

2.4. Since

$$\mathcal{E}_n^{-1}(A) = \left(\prod \mathbf{E}(\exp(\alpha_k \Delta V_k) \mid \mathcal{F}_{k-1})\right)^{-1},$$

it follows that

$$Z_n = \exp\left(\sum_{k=1}^{n} \alpha_k \Delta V_k\right)\left(\prod \mathbf{E}(\exp(\alpha_k \Delta V_k) \mid \mathcal{F}_{k-1})\right)^{-1}$$

$$= \prod_{k=1}^{n} \exp(\alpha_k \Delta V_k)(\mathbf{E}(\exp(\alpha_k \Delta V_k) \mid \mathcal{F}_{k-1}))^{-1},$$

and hence

$$\mathbf{E}(Z_n \mid \mathcal{F}_{n-1}) = Z_{n-1}\mathbf{E}(\exp(\alpha_n \Delta V_n)(\mathbf{E}(\exp(\alpha_n \Delta V_n) \mid \mathcal{F}_{n-1}))^{-1} \mid \mathcal{F}_{n-1}) = Z_{n-1}.$$

3.1. We represent the stochastic sequence V_n in (3.2) according to the Doob decomposition in the form $V_n = M_n + A_n$, where M is a martingale and A is a predictable sequence. By Girsanov's theorem, the stochastic sequence

$$M_n^* = (V - A)_n^* = (V - A)_n - \sum_{k=1}^{n} \mathbf{E}(Z_{k-1}^{-1} Z_k \Delta (V - A)_k \mid \mathcal{F}_{k-1})$$

is a martingale with respect to the measure $\mathbf{P}^*$ with local density Z_n with respect to $\mathbf{P}$. In view of Theorem 2.1 the density Z must be chosen so that

$$r_n = \Delta U_n = \Delta A_n + \mathbf{E}(\mathbf{E}(Z_{n-1}^{-1} Z_n \Delta (V - A)_n \mid \mathcal{F}_{n-1})$$
$$= \Delta A_n + \mathbf{E}(Z_{n-1}^{-1} Z_n \Delta V_n \mid \mathcal{F}_{n-1}) - Z_{n-1}^{-1} \Delta A_n \mathbf{E}(Z_n \mid \mathcal{F}_{n-1})$$
$$= \mathbf{E}(Z_{n-1}^{-1} Z_n \Delta V_n \mid \mathcal{F}_{n-1}) = \mathbf{E}(Z_{n-1}^{-1} Z_n \rho_n \mid \mathcal{F}_{n-1}),$$

that is, the formula connecting r, ρ, and Z is preserved for the "martingale" case.

3.2. To construct a spatially infinite market $(d = \infty)$ we consider the space $\mathbf{R}^{\mathbb{Z}+}$ of numerical sequences. Let $\Omega = \mathbb{Z}_+$, $\mathcal{F}_0 = \{\varnothing, \Omega\}$, $\mathcal{F}_1 = \mathcal{F} = 2^\Omega$, and $\mathbf{P} = \sum_{n=1}^{\infty} 2^{-n} \delta_n$, where $\delta_n\{k\} = 1$ for $k = n$ and 0 for $k \neq n$. We define an $\mathbf{R}^{\mathbb{Z}+}$-valued stochastic sequence $(S_n)_{n=0,1}$ by

$$S_0 \equiv 0, \quad S_1^j(\omega) = \begin{cases} 1 & \text{for } \omega = j, \\ -1 & \text{for } \omega = j+1, \\ 0 & \text{otherwise,} \end{cases}$$

where S^j is the jth coordinate. Then the relation $\mathbf{P}^*\{j\} = \mathbf{P}^*\{j+1\}$ must hold for a martingale measure $\mathbf{P}^*$, which cannot be true for arbitrary $j \in \mathbb{Z}_+$. Consequently, such a measure does not exist.

For the other "infinite" case $(N = \infty)$ we follow [20] and define a "risky" asset $S = (S_k)_{k \in \mathbb{Z}_+}$ to be the sequence of partial sums of independent random variables

$$\xi_k = \begin{cases} 1 & \text{with probability } p \neq 1/2, \\ -1 & \text{with probability } 1 - p, \end{cases}$$

and we let $\mathcal{F}_k = \sigma(\xi_1, \ldots, \xi_k)$. Now if $\mathbf{P}^*$ is a martingale measure, then $\mathbf{P}^*(\xi_k = 1 \mid \mathcal{F}_{k-1}) = 1/2$ (a.s.). Further, by the law of large numbers for the measures $\mathbf{P}$ and $\mathbf{P}^*$, $S_n/n \to 2p - 1 \neq 0$ $\mathbf{P}$-a.s. and $S_n/n \to 0$ $\mathbf{P}^*$-a.s. as $n \to \infty$, and this means that $\mathbf{P}$ and $\mathbf{P}^*$ are not equivalent.

3.3. The assertions a)–c) are simple exercises with conditional expectations. From them,

$$X_0^N = \mathbf{E} X_N^N Z_N \geq \mathbf{E}\{X_N^N Z_N \mathbf{I}_{\{-C_N \leq X_N^N < 0\}}\} + \mathbf{E}\{X_N^N Z_N \mathbf{I}_{\{X_N^N \geq \varepsilon\}} \mathbf{I}_{\{Z_N \geq \varepsilon\}}\}$$
$$\geq -C_N + \varepsilon^2 \mathbf{P}\{X_N^N \geq \varepsilon, \ Z_N \geq \varepsilon\}.$$

Further,

$$X_0^N + C_N \geq \varepsilon \mathbf{P}\{X_N^N \geq \varepsilon, \ Z_N \geq \varepsilon\} \geq \varepsilon^2 (\mathbf{P}\{X_N^N \geq \varepsilon\} - \mathbf{P}\{Z_N < 0\}).$$

Passing to the limit, we get that $\mathbf{P}\{Z_\infty < \varepsilon\} \geq \limsup_{N \to \infty} \mathbf{P}\{X_N^N \geq \varepsilon\}$. If the portfolio π^{*N} is asymptotically arbitrage-free, then there exists an $\varepsilon' > 0$ such that $\limsup_{N \to \infty} \mathbf{P}\{X_N^{*N} \geq \varepsilon'\} = \delta$ for some $\delta > 0$. Consequently, $\mathbf{P}\{Z_\infty < \varepsilon\} \geq \delta$ for $\varepsilon > \varepsilon'$, which contradicts a condition of the problem. We remark that the condition $\mathbf{P}\{Z_\infty < 0\} = 1$ means the continuity of $\{\mathbf{P}^N\}$ with respect to $\{\mathbf{P}^{*N}\}$ in the sense that if $A^N \in \mathcal{F}_N$ and if $\mathbf{P}^{*N}(A^N) \to 0$ as $N \to \infty$, then $\mathbf{P}^N(A^N) \to 0$ as $N \to \infty$.

4.1. For a complete no-arbitrage $(1, S)$-market (3.1) Theorem 4.1 and Jensen's inequality give us that

$$\begin{aligned}
\mathbb{C}_{(N+1)} &= \mathbf{E}^*((S_{N+1} - K)^+) = \mathbf{E}^*((S_N(1 + \rho_{N+1}) - K)^+) \\
&= \mathbf{E}^*(\mathbf{E}^*((S_{N+1} - K)^+) \mid \mathcal{F}_N) \\
&\geq \mathbf{E}^*((S_N \mathbf{E}^*((1 + \rho_{N+1}) \mid \mathcal{F}_N) - K)^+) \\
&= \mathbf{E}^*((S_N - K)^+) = \mathbb{C}_{(N)}.
\end{aligned}$$

4.2. The cases a) and c) are simple exercises, so we dwell on the more complicated case c) for a complete $(1, S)$-market (3.2) with a unique martingale measure $\mathbf{P}^*$. Since the function $(x - K)^+$ is convex with respect to K, we have for $p \in [0, 1]$ that

$$\begin{aligned}
p(S_0 \mathcal{E}_N(V) &- K_1)^+ + (1 - p)(S_0 \mathcal{E}_N(V) - K_2)^+ \\
&= \mathcal{E}_N(V)(p(S_0 - K_1 \mathcal{E}_N^{-1}(V))^+ + (1 - p)(S_0 - K \mathcal{E}_N^{-1}(V))^+) \\
&\geq \mathcal{E}_N(V)(S_0 - pK_1 \mathcal{E}_N^{-1}(V) - (1 - p)K_2 \mathcal{E}_N^{-1}(V))^+ \\
&= (S_0 \mathcal{E}_N(V) - pK_1 - (1 - p)K_2)^+.
\end{aligned}$$

Taking the mathematical expectation with respect to the measure $\mathbf{P}^*$ now gives the necessary inequality:

$$p\mathbb{C}(S_0, N, K_1) + (1 - p)\mathbb{C}(S_0, N, K_2) \geq \mathbb{C}(S_0, N, pK_1 + (1 - p)K_2).$$

Convexity with respect to S_0 is proved similarly.

4.3. Let π_1 and π_2 be two minimal hedges on a complete (B, S)-market (3.1) with martingale measure $\mathbf{P}^*$. Then with regard to this measure the sequences

$$M_n^i = B_n^{-1} X_n^{\pi^i} = \mathbb{C}(N) B_0^{-1} + \sum_{k=1}^n B_k^{-1} \gamma_k^i S_{k-1}(\rho_k - r_k), \qquad i = 1, 2,$$

are martingales, and $M_N^1 = B_N^{-1} f = M_N^2$. Their difference

$$M_n = M_n^1 - M_n^2 = \sum_{k=1}^n B_k^{-1}(\gamma_k^1 - \gamma_k^2) S_{k-1}(\rho_k - r_k)$$

is also a martingale such that $M_n = \mathbf{E}^*(M_N \mid \mathcal{F}_n)$ and $M_0 = M_N = 0$. Consequently, $M_n \equiv 0$, and hence $\gamma_k^1 = \gamma_k^2$, $k \leq N$.

4.4. It is clear that the required exchange rate is equal to the ratio S_n^1/S_n^2, and thus is a solution of the difference equation $\Delta(S^1/S^2)_n = (S_0^1/S_0^2)(\rho_n^1 - \rho_n^2)/(1+\rho_n^2)$.

5.1. Using the procedure described in the proof of Theorem 5.1 for constructing $\tau^* = \tau_1^* = \min\{0 \leq k \leq 1 : Y_k = f_k/B_k\}$ and $\mathbb{C}(1) = \sup_{0 \leq \tau \leq 1} \mathbf{E}^*(1 + r)^{-1}\beta^*(S_\tau - 1)^+$, we get

a) $\frac{f_0}{B_0} = \frac{(S_0 - 1)^+}{1} = 0 < \mathbf{E}^*Y_1 = \alpha\beta p^*(\lambda - 1) \implies \tau^* \equiv 1$, and $\mathbb{C}(1) = \mathbf{E}^*Y_1 = \alpha\beta p^*(\lambda - 1)$;

b) $\frac{f_0}{B_0} = \lambda - 1$, $\mathbf{E}^*Y_1 = \alpha\beta p^*(\lambda^2 - 1) = (1 + r)^{-1}\beta(\lambda^2 - 1)\frac{r - (\lambda^{-1} - 1)}{\lambda - 1 - (\lambda^{-1} - 1)} = \beta(\lambda - \alpha)$, and if $\beta > \frac{\lambda - 1}{\lambda - \alpha}$, then $\mathbf{E}^*Y_1 > \lambda - 1 \implies \tau^* \equiv 1$ and $\mathbb{C}(1) = \mathbf{E}^*Y_1 = \beta(\lambda - \alpha)$.

5.2. As in the solution of Problem 5.1, to find the optimal stopping time τ^* we must compare the quantities $f_0/B_0 = \lambda^k - 1$ and $\mathbf{E}^*Y_1 = \mathbf{E}^*(S_1 - 1)^+\alpha\beta = \alpha\beta p^*(\lambda^{k+1} - 1) + \alpha\beta(1 - p^*)(\lambda^{k-1} - 1) = \alpha\beta(p^*(\lambda^{k+1} - \lambda^{k-1}) + \lambda^{k-1} - 1) = \beta(\lambda^k - \alpha)$. Thus, $\tau_1^* = 1$ for $\lambda^k - 1 < \beta(\lambda^k - \alpha)$, and otherwise $\tau_1^* = 0$. Correspondingly,

$$\mathbb{C}(1) = \sup_{0 \leq \tau \leq 1} \mathbf{E}^*(1 + r)^{-\tau^*}\beta^{\tau^*}(S_{\tau^*} - 1)^+ = \max(\lambda^k - 1, \beta(\lambda^k - \alpha)).$$

These arguments lead to the solution.

6.1. The solution here is analogous to 4.3.

6.2. From (6.6) and (6.7),

$$\mathbf{E}^*\mathcal{E}_N^{-1}X_N^{\Pi(G)} = X_0^{\Pi(G)} - \sum_{k=1}^{N}\mathcal{E}_{k-1}^{-1}\mathbf{E}^*\Delta G_k \leq X_0^{\Pi(G)} - \sum_{k=1}^{N}\mathcal{E}_{k-1}^{-1}\mathbf{E}^*\Delta D_k.$$

Hence, $X_0^{\Pi(G)} \geq \mathbf{E}^*\mathcal{E}_N^{-1}X_N^{\Pi(G)} + \sum_{k=1}^{N}\mathcal{E}_{k-1}^{-1}\mathbf{E}^*\Delta D_k$. In particular, for $\Delta G_n = \mathbb{C}_G S_{n-1}$ and $\Delta D_n = \mathbb{C}_D S_{n-1}$ the resulting inequality is valid when $\mathbb{C}_G \geq \mathbb{C}_D$.

7.1. Suppose that on a $(1, S)$-market the original measure $\mathbf{P} = \mathbf{P}^*$ is a martingale measure. Minimizing the remaining risk

$$R_n^{\Pi} = \mathbf{E}\left(\left(f - X_n^{\Pi} - \sum_{k=1}^{N}\gamma_k S_{k-1}\rho_k\right)^2 \mid \mathcal{F}_n\right)$$

with respect to X and γ, we get the equations

$$\frac{\partial R_{N-1}^{\Pi}}{\partial X_{N-1}} = 2\mathbf{E}(f - X_{N-1}^{\Pi} - \gamma_N S_{N-1}\rho_N \mid \mathcal{F}_{N-1}) = 0,$$

$$\frac{\partial R_{N-1}^{\Pi}}{\partial \gamma_{N-1}} = 2\mathbf{E}((f - X_{N-1}^{\Pi} - \gamma_N S_{N-1}\rho_N)S_{N-1}\rho_N \mid \mathcal{F}_{N-1}) = 0.$$

Therefore, the risk-minimizing strategy is determined by the relations $\widetilde{X}_{N-1} = \mathbf{E}(f \mid \mathcal{F}_{N-1})$ and $\widetilde{\gamma}_N = \mathbf{E}(f\rho_N \mid \mathcal{F}_{N-1})/(S_{N-1}\mathbf{E}(\rho_N^2 \mid \mathcal{F}_{N-1}))$ (the procedure is extended in the natural way to $k < N$: $\widetilde{\gamma}_k = \mathbf{E}(f\rho_k \mid \mathcal{F}_{k-1})/(S_{k-1}\mathbf{E}(\rho_k^2 \mid \mathcal{F}_{k-1}))$, $\widetilde{X}_k = \mathbf{E}(f \mid \mathcal{F}_{k-1})$). For an optimal strategy $\widehat{\pi}$ it follows from the martingale property of $G_n = \sum_{k=1}^{n}\Delta G_k$ and $M_n = \sum_{k=1}^{n}\rho_k$ that $\widehat{X}_n = \mathbf{E}(f \mid \mathcal{F}_n)$. Since M

and G are orthogonal, it follows that $0 = \mathbf{E}(\rho_k \Delta G_k \mid \mathcal{F}_{k-1}) = -\mathbf{E}(\rho_k \Delta \widehat{X}_k \mid \mathcal{F}_{k-1}) + \widehat{\gamma}_k \mathbf{E}(\rho_k \Delta S_k \mid \mathcal{F}_{k-1})$ or $\widehat{\gamma}_k = \mathbf{E}(f \rho_k \mid \mathcal{F}_{k-1})/(S_{k-1}\mathbf{E}(\rho_k^2 \mid \mathcal{F}_{k-1}))$. Comparison of the formulas obtained leads to the necessary assertion of the problem.

7.2. For a self-financing portfolio $\pi_n = (\beta_n, \gamma_n)$ the value X_n^π is determined by the formula $(3.5')$. Using $(3.5')$ and the notation $X_n^\gamma = \mathcal{E}_n^{-1} X_n^\pi$, $f_N = \mathcal{E}_n^{-1} f$, $X_0^\pi = x$, and $\Delta S_n^1 = \mathcal{E}_n^{-1} S_{n-1}(\rho_n - r_n)$, we get that $X_n^\gamma = X_n^\gamma(x) = x + \sum_{k=1}^n \gamma_k \Delta S_k^1$. Hence, the original problem reduces to finding an $x^* > 0$ and a predictable sequence γ^* such that

$$R_N^{\gamma^*}(x^*) \equiv \mathbf{E}(X_N^{\gamma^*}(x^*) - f_N)^2 = \inf_{x,\gamma} R_N^\gamma(x).$$

Here the first component β^* of the optimal self-financing portfolio π^* can be uniquely recovered from γ^*.

Note also that (S_n^1) and $(X_n^\gamma(x))$ are martingales, and $x = \mathbf{E} X_n^\gamma(x)$. Further, rewriting $R_N^\gamma(x) = (\mathbf{E}(f_N - x))^2 + \mathbf{E}((X_N^\gamma(x) - x) - (f_N - \mathbf{E} f_N))^2$ and differentiating with respect to x, we have that $(R_N^\gamma(x))' = -2\mathbf{E}(f_N - x) = 0$. Consequently, $x^* = \mathbf{E} f_N$. Now take hypothetical optimal variables to be $\gamma_n^* = \mathbf{E}(f_N \Delta S_n^1 \mid \mathcal{F}_{n-1})/\mathbf{E}((\Delta S_n^1)^2 \mid \mathcal{F}_{n-1})$, and form the martingale

$$L_n^* = \mathbf{E}(f_N \mid \mathcal{F}_n) - \mathbf{E}\Big(\sum \gamma_k^* \Delta S_k^1 \,\Big|\, \mathcal{F}_n\Big) - x^*.$$

Setting $n = N$, we get, for example, the representation

$$f_N = x^* + \sum_{k=1}^N \gamma_k^* \Delta S_k^1 + L_N^*.$$

Since the martingales $\sum_{k=1}^N \gamma_k^* \Delta S_k^1$ and L_N^* are orthogonal,

$$\mathbf{E}(\Delta L_n^* \gamma_n^* \Delta S_n^1 \mid \mathcal{F}_{n-1}) = 0.$$

Next,

$$R_N^\gamma(x^*) = \mathbf{E}\Big(f_N - \Big(x^* + \sum_{k=1}^N \gamma_k^* \Delta S_k^1\Big)\Big)^2 = \mathbf{E}\Big(\sum_{k=1}^N (\gamma_k^* - \gamma_k)\Delta S_k^1 + L_N^*\Big)^2$$

$$= \mathbf{E}\Big(\sum_{k=1}^N (\gamma_k^* - \gamma_k)\Delta S_k^1\Big)^2 + \mathbf{E}(L_N^*)^2 \geq \mathbf{E}(L_N^*)^2.$$

It is clear that equality here is possible only when $\gamma_k = \gamma_k^*$.

7.3. Suppose that an option has cost $x > \mathbb{C}^*$, and it was bought at this price. Then the seller has the following arbitrage opportunity. For this it is necessary to invest capital $y \in (\mathbb{C}^*, x)$ on a (B, S)-market in such a way that $X_0^{\pi^*}(y) = y$ and $X_N^{\pi^*}(y) \geq f$. This is possible in view of the definition of $\mathbb{C}^*$. Here the nonrisky gain of the seller is strictly positive:

$$(x - f) + (X_N^{\pi^*}(y) - y) = (x - y) + (X_N^{\pi^*}(y) - f) \geq x - y > 0.$$

But if the buyer acquired the contract at a price $x < \mathbb{C}^*$, then an arbitrage opportunity is realized by the buyer. According to the definition of $\mathbb{C}^*$, there exists for $y \in (\mathbb{C}_*, x)$ a portfolio π^* such that

$$X_0^{\pi^*}(y) = y, \qquad X_N^{\pi^*}(y) \leq f.$$

Here the buyer proceeds as follows: he borrows y and invests it according to the portfolio π^* and receives the nonrisky profit

$$(f - X_N^{\pi^*}(y)) + (y - x) \geq y - x > 0.$$

7.4. There is a hint in the formulation of the problem.

8.1. Since the total expenditures of the parties in a forward contract are zero, for the contract to be represented as a composition of a call option and a put option it is necessary that these options have equal prices: $\mathbb{C}_N = \mathbb{P}_N$. Further, it follows from the call-put parity $\mathbb{C}_N = \mathbb{P}_N + S_0 - K(1+r)^{-N}$ that $K = S_0(1+r)^N$. The latter quantity coincides precisely with the strike price for the forward contract.

8.2. A forward contract (on an asset $(S_n)_{n \leq N}$) with strike price F_n is equivalent to accepting the contingent claim

$$f_k = \begin{cases} 0, & k = n, \ldots, N-1, \\ S_N - F_n, & n = N. \end{cases}$$

It follows from the general theory of pricing contingent claims that a no-arbitrage price $\mathbb{C}_n^F$ must be found within the limits

$$\mathbb{C}_n = \inf_{\tilde{\mathbf{P}} \in \mathbb{P}^*} \widetilde{\mathbf{E}}(S_N - F_n \mid \mathcal{F}_n) \leq \mathbb{C}_n^F \leq \sup_{\tilde{\mathbf{P}} \in \mathbb{P}^*} \widetilde{\mathbf{E}}(S_N - F_n \mid \mathcal{F}_n) = \mathbb{C}^n.$$

On the other hand, the cost $\mathbb{C}_n^F$ for the forward contract is equal to zero. Consequently, $\mathbb{C}_n \leq 0$ and $\mathbb{C}^n \geq 0$, which implies the required assertion.

8.3. To determine the futures prices F_n^* on the complete (B, S)-market with martingale measure $\mathbf{P}^*$ we use "backward induction". In concluding the futures contract at time $N-1$ for the purchase of a unit of the asset at the price F_{N-1}^* it is necessary to deposit in the margin account $\Delta M_n = m_n M_{n-1}$, $-1 < m_n \leq r_n$, the quantity

$$\alpha_N F_{N-1}^* = \mu_N M_{N-1},$$

where μ_N is the number of units in this account. At time N the investor gets

$$S_N - F_{N-1}^* + \mu_N M_N.$$

Consequently, the contingent claim

$$f_N = S_N - F_N^* + \mu_N M_N$$

is connected with this futures contract, and its cost is

$$\mathbb{C}_{N-1} = \mathbf{E}^* \left(\frac{B_{N-1}}{B_N} (S_N - F_{N-1}^* + \mu_N M_N) \,\middle|\, \mathcal{F}_{N-1} \right).$$

Since there are no arbitrage opportunities,

$$\alpha_N F_{N-1}^* \geq \mathbf{E}^* \left(\frac{B_{N-1}}{B_N} (S_N - F_{N-1}^* + \mu_N M_N) \,\middle|\, \mathcal{F}_{N-1} \right),$$

and hence

$$F_{N-1}^* \geq \frac{F_{N-1}}{1 + \alpha_N (r_N - m_N)}.$$

Conclusion of the futures contract to sell is equivalent to the contingent claim

$$f_N' = F_{N-1}^* - S_N + \mu_N M_N, \qquad \alpha_N F_{N-1}^* = \mu_N M_{N-1}.$$

In a way analogous to that above we get that

$$F^*_{N-1} \leq \frac{F_{N-1}}{1 - \alpha_N(r_N - m_N)}.$$

Continuing these arguments leads to the establishment of bounds for the futures prices F^*_n for all $n \leq N$.

8.4. This can be established directly.

9.1. Let $B_0 = B_1$. Then the yield of the portfolio π takes the form $R = R^\pi = X^\pi_1/X_0 - 1$, and maximization of the mean yield $\mathbf{E}R^\pi$ reduces to maximization of the expectation

$$\mathbf{E}X^\pi_1 = X_0 + \gamma_1 S_0(bp + a(1 - p)) - \delta S_0|\Delta\gamma_1|$$

with respect to γ_1.

The function obtained is piecewise linear in γ_1 and attains its maximum at the break point $\gamma_1 = \gamma_0$ (the portfolio remains as before) or on the limits $\gamma_1 = 0$ (all the stocks are sold at time zero) and $\gamma_1 = (X_0 - \gamma_0 S_0)/(S_0(1 + \delta)) + \gamma_0$ of the possible values. The rest is a direct computation of $\mathbf{E}X_1$ and $\sup_\pi \mathbf{E}R^\pi$.

9.2. For the single-period model of a $(1, S)$-market it follows since γ_1 is deterministic (γ_1 is measurable with respect to $\mathcal{F}_0 = \{\varnothing, \Omega\}$) that the problem reduces to the standard problem of minimizing the quadratic (with respect to γ_1) function $\mathbf{E}(X^\pi_1 - f)^2$ in the two regions $\gamma_1 \geq \gamma_0$ and $\gamma_1 < \gamma_0$.

9.3. It is clear that for the function $u(x) = x^\alpha/\alpha$ the corresponding function v is $v(x) = x^{1/(\alpha-1)}$, and the form of the optimal terminal value is $X^*_N = (yZ^*_N)^{1/(\alpha-1)}$. Further,

$$\psi(y) = \mathbf{E} \sup_{X_N>0} (X^\alpha_N/\alpha - yZ^*_N X_N)$$

$$= \mathbf{E}((X^*_N)^\alpha/\alpha - (yZ^*_N)X^*_N) = \left(\frac{1}{\alpha} - 1\right)y^{\alpha/(\alpha-1)}\mathbf{E}(Z^*_N)^{\alpha/(\alpha-1)},$$

and, with use of the notation $c = \mathbf{E}(Z^*_N)^{\alpha/(\alpha-1)}$,

$$\varphi(x) = \inf_{y>0}(\psi(y) + yx) = \inf_{y>0}(c(1/\alpha - 1)y^{\alpha/(\alpha-1)} + yx).$$

The quantity $\hat{y} = (x/c)^{\alpha-1}$ minimizes the function $c(1/\alpha - 1)y^{\alpha/(\alpha-1)} + yx$, and hence

$$\varphi(x) = \left(\frac{1}{\alpha} - 1\right)\frac{x^\alpha}{c^{\alpha-1}} + \frac{x^\alpha}{c^{\alpha-1}} = \frac{x^\alpha}{\alpha c^{\alpha-1}}, \quad X^*_N = \frac{x}{c}(Z^*_N)^{1/(\alpha-1)}.$$

To find an optimal strategy $\pi^*_n = (\beta^*_n, \gamma^*_n)$ we introduce the proportion α^*_n of the risky part of the portfolio. Equating the two expressions for the terminal value gives us that

$$\mathcal{E}_N\left(\sum \alpha^*_k \rho_k\right) = c^{-1}\mathcal{E}_N^{1/(\alpha-1)}\left(\sum\left(-\frac{m}{d}(\rho_k - m)\right)\right),$$

and, as a consequence,

$$\alpha^*_k = \frac{1}{b}\left\{\left(\frac{a}{a-b} + \left(\frac{-bp}{a(1-p)}\right)^{1/(\alpha-1)}\frac{b}{b-a}\right)^{-1} - 1\right\}.$$

As a result, π_k^* is found from the formulas

$$\beta_k^* = X_{k-1}^*(1 - \alpha_k^*), \quad \gamma_k^* = \alpha_k^* X_{k-1}^*/S_{k-1}.$$

10.1. Use the Black–Scholes formula directly.

10.2. Considering a (B, S, Δ)-market with parameters satisfying (10.6), we find that a martingale measure satisfies

$$p^* = \frac{(e^{r\Delta} - 1) - (1 - e^{-\sigma\sqrt{\Delta}})}{e^{\sigma\sqrt{\Delta}} - e^{-\sigma\sqrt{\Delta}}} = \frac{1}{2}\left(1 + \frac{r}{\sigma}\sqrt{\Delta}\right) + o(\Delta).$$

Further, since this is a "binomial" model,

$$e^{r\Delta}\mathbb{C}(s, t) = p^*\mathbb{C}(Se^{\sigma\sqrt{\Delta}}, t + \Delta) + (1 - p^*)\mathbb{C}(Se^{-\sigma\sqrt{\Delta}}, t + \Delta).$$

From Taylor's formula we get that as $\Delta \to 0$

$$e^{r\Delta}\mathbb{C}(s, t) = (1 + r\Delta)\mathbb{C}(s, t) + o(\Delta);$$

$$\mathbb{C}(Se^{\sigma\sqrt{\Delta}}, t + \Delta) = \mathbb{C}(s, t) + \frac{\partial}{\partial t}\mathbb{C}(s, t) + \frac{\partial\mathbb{C}}{\partial s}s\sigma\sqrt{\Delta} + \frac{1}{2}\frac{\partial^2\mathbb{C}}{\partial s^2}s^2\sigma^2\Delta + o(\Delta);$$

$$\mathbb{C}(Se^{-\sigma\sqrt{\Delta}}, t + \Delta) = \mathbb{C}(s, t) + \frac{\partial}{\partial t}\mathbb{C}(s, t) - \frac{\partial\mathbb{C}}{\partial s}s\sigma\sqrt{\Delta} + \frac{1}{2}\frac{\partial^2\mathbb{C}}{\partial s^2}s^2\sigma^2\Delta + o(\Delta).$$

Consequently,

$$(1 + r\Delta)\mathbb{C}(s, t) = \mathbb{C}(s, t) + \frac{\partial\mathbb{C}}{\partial t}\Delta + \frac{\partial\mathbb{C}}{\partial s}rs\Delta + \frac{1}{2}\frac{\partial^2\mathbb{C}}{\partial s^2}s^2\sigma^2\Delta + o(\Delta),$$

and passage to the limit as $\Delta \to 0$ leads to the Black–Scholes equation.

Bibliography

1. A. N. Burenin, *Futures, forward, and option contracts*, Trivola, Moscow, 1995. (Russian)
2. J. Jacod and A. N. Shiryaev, *Limit theorems for stochastic processes*, Springer–Verlag, Berlin, 1987; Russian transl., vols. 1, 2, Fizmatlit, 1994.
3. D. O. Kramkov and A. N. Shiryaev, *On rational pricing of a "Russian option" in the symmetric binomial model of a (B, S)-market*, Teor. Veroyatnost. i Primenen. **39** (1994), 191–200; English transl. in Theory Probab. Appl. **39** (1994).
4. A. V. Mel′nikov and A. N. Shiryaev, *Criteria for the absence of arbitrage in the financial market*, Progress in the Theory of Probability and its Applications II (A. N. Shiryaev, et al., eds.), TVP, Moscow, 1996, pp. 121–134.
5. A. V. Mel′nikov, M. L. Nechaev, and V. M. Stepanov, *On a discrete model for a financial market and methods for pricing securities*, preprint no. 3, Actuarial and Financial Center for Scientific Investigation, Moscow, 1996. (Russian)
6. A. A. Novikov, *Pricing options of American type: a minimax statistical approach*, Teor. Veroyatnost. i Primenen. (to appear); English transl. in Theory Probab. Appl..
7. A. A. Pervozvanskiĭ and T. N. Pervozvanskaya, *Financial markets: pricing and risk*, Infra-M, Moscow, 1994. (Russian)
8. S. T. Rachev and L. Rüschendorf, *Models for option prices*, Teor. Veroyatnost. i Primenen. **39** (1994), 150–190; English transl. in Theory Probab. Appl. **39** (1994).
9. H. Robbins, D. Siegmund, and Y. S. Chow, *Great expectations: The theory of optimal stopping*, Houghton–Mifflin, Boston, 1971; Russian transl., "Nauka", Moscow, 1977.
10. A. N. Shiryaev, *Probability*, 2nd ed., "Nauka", Moscow, 1989; English transl., Springer–Verlag, Berlin–New York, 1996.
11. A. N. Shiryaev, *On some basic concepts and stochastic models in financial mathematics*, Teor. Veroyatnost. i Primenen. **39** (1994), 5–22; English transl. in Theory Probab. Appl. **39** (1994).
12. A. N. Shiryaev, Yu. M. Kabanov, D. O. Kramkov, and A. V. Mel′nikov, *Toward a theory of pricing options of European and American types. I. Discrete time*, Teor. Veroyatnost. i Primenen. **39** (1994), 23–79; English transl. in Theory Probab. Appl. **39** (1994).
13. L. Bachelier, *Théorie de la spéculation*, Ann. École Norm. Sup. **17** (1900), 21–86; reprint, The Random Character of Stock Market Prices (P. H. Coothner, ed.), MIT Press, Cambridge, MA, 1967, pp. 17–78.
14. J. Bardhan and X. Chao, *Pricing options on securities with discontinuous returns*, Stochastic Process. Appl. **48** (1993), no. 1, 123–137.
15. F. Black and M. Scholes, *The pricing of options and corporate liabilities*, J. Polit. Economy **3** (1973), 637–659.
16. D. B. Colwell and R. J. Elliot, *Discontinuous asset prices and non-attainable contingent claims and corporate policy*, Math. Finance **3** (1993), no. 3, 295–368.
17. T. Copeland and J. Westen, *Financial Theory and Corporate Policy*, Addison–Wesley, Reading, MA, 1983.
18. J. C. Cox and M. Rubinstein, *Option Markets*, Prentice–Hall, Englewood Cliffs, NJ, 1985.
19. J. C. Cox, R. A. Ross, and M. Rubinstein, *Option pricing: a simplified approach*, J. Financial Econom. **3** (1979), no. 7, 229–263.
20. R. C. Dalang, A. Morton, and W. Willinger, *Equivalent martingale measures and no-arbitrage in stochastic securities market models*, Stochastics Stochastics Rep. **29** (1990), no. 2, 185–209.
21. R. A. Dana and M. Jeanblanc-Picqué, *Marchés financiers en temps continu (valorisation et équilibre)*, Economica, Paris, 1994.
22. F. Delbaen and W. Schachermayer, *A general version of the fundamental theorem of asset pricing*, Math. Ann. **300** (1994), 463–520.

23. M. V. Dothan, *Prices in Financial Markets*, Oxford Univ. Press, Oxford, 1990.

24. D. Duffie, *Dynamic Asset Pricing Theory*, Princeton Univ. Press, Princeton, NJ, 1992.

25. H. Föllmer and D. Sonderman, *Hedging of non-redundant contingent claims*, Contributions to Mathematical Economics (W. Hildebrand and A. Mas-Colell, eds.), 1986, pp. 205-223.

26. H. Föllmer, *Probabilistic aspects of options*, Discussion paper B-202, Universität Bonn, Bonn, 1991.

27. H. Föllmer and M. Schweizer, *Hedging of contingent claims under incomplete information*, Applied Stochastic Analysis (M. N. A. Davis and R. J. Elliott, eds.), Gordon & Breach, London, 1991, pp. 389–408.

28. J. M. Harrison and D. M. Kreps, *Martingales and arbitrage in multiperiod securities markets*, J. Econom. Theory **20** (1979), 381–408.

29. J. M. Harrison and S. R. Pliska, *Martingales and stochastic integrals in the theory of continuous trading*, Stochastic Process. Appl. **11** (1981), no. 3, 215–260.

30. T. S. Y. Ho and Sang-Bin Lee, *Term structure movements and pricing interest rate contingent claims*, J. Finance **41** (1986), 1011–1029.

31. J. Hull, *Options, futures and other derivative securities*, Prentice-Hall, Englewood Cliffs, NJ, 1992.

32. I. Karatzas, J. P. Lechoczky, S. E. Shreve, and G. L. Xu, *Martingale and duality methods for utility maximization in an incomplete market*, SIAM J. Control Optim. **29** (1991), 702–730.

33. D. O. Kramkov, *Optional decomposition of supermartingales and hedging continent claims in incomplete security markets*, Probab. Theory Related Fields **105** (1996), 459–479.

34. H. Markowitz, *Mean-variance analysis in portfolio choice and capital markets*, Blackwell, Cambridge, MA, 1990.

35. R. Merton, *Option pricing when underlying stock returns are discontinuous*, J. Finan. Econom. **3** (1976), 125–144.

36. R. Merton, *Continuous-time finance*, Blackwell, Cambridge, MA, 1993.

37. W. Schachermayer, *A Hilbert space proof of the fundamental theorem of asset pricing in finite discrete time*, Insurance: Math. Econom. **11** (1992), 249–257.

38. W. Schachermayer, *A counterexample to several problems in the theory of asset pricing*, Math. Finance **3** (1993), no. 2, 217–229.

39. M. Schweizer, *Approximation pricing and the variance-optimal martingale measure*, Ann. Probab. **24** (1996), no. 1, 206–236.

40. M. S. Taqqu and W. Willinger, *The analysis of finite security markets using martingales*, Adv. Appl. Probab. **19** (1987), 1–25.

Subject Index

account
— , bank 5
— , margin 74
actuarial mathematics 19
appreciation rate 94
arbitrage (arbitrage opportunity)
    4, 22
— , asymptotic 30
asset 1, 2, 4
— , nonrisky 2, 21
— , risky 21
best linear estimator 71
binomial lattice 99
bond 2, 79
— , convertible 3
— with discount (zero-coupon
    bond) 79
broker 6
call-put parity 41
cap 4
clearing house 6
completeness of a market 27, 63,
consumption 55
contingent claim 31, 47, 65
— — , attainable 32
— — , attainable in probability
    42
— — , dynamic 47
coupon 2, 79
covariance 11
debut of a set 8, 15
decomposition
— , Doob 10
— , Kunita-Watanabe 11
— , multiplicative 16
density 9
— , local 10, 13
dividend 58, 61

duration 81
— analysis 81
— , Macaulay 81
face value (par value, principal) 2,
    79
filtration 7
floor 4
formula
— , Black-Scholes 95, 96
— , Itô's 17
— , Cox-Ross-Rubinstein 40
forward (contract) 73
fundamental theorem of financial
    mathematics 23
futures (contract) 74
hedge
— , α- 42
— , $(\mathbb{C}, f, N)$- 33
— , G- 56
— , (x, f, N)- 32, 47
— , minimal 32, 33, 40, 48
hedging 4, 31
insurance 2
investment
— cost 32
— goal 78
investor 4
leasing 99
loan 5
maker 6
margin 6, 74
market
— , (B, S)- 22, 31, 35
— , complete 27, 63
— , currency 41
— , financial 1, 21, 31
— , incomplete 63

Selected Titles in This Series

(Continued from the front of this publication)

(See the AMS catalog for earlier titles)